CHRISTOPHER SAXTON'S
16th CENTURY MAPS

CHRISTOPHER SAXTON'S
16th CENTURY MAPS

The Counties of England & Wales

With an Introduction

by

WILLIAM RAVENHILL

*Emeritus Reardon Smith Professor of Geography
at University of Exeter*

CHATSWORTH
LIBRARY

First published in the UK in 1992 by Chatsworth Library
an imprint of Airlife Publishing Ltd.

British Library Cataloguing in Publication Data
A catalogue record for this book
is available from the British Library

ISBN 1 85310 354 3

Printed by Kyodo Printing Company (S'pore) Pte. Ltd.

Chatsworth Library

an imprint of Airlife Publishing Ltd.
101 Longden Road, Shrewsbury SY3 9EB

Clemens et Regni moderatrix iusta Britāni
Hac forma insigni conspicienda nitet.

Tristia dum gentes circum omnes bella fatigant;
Cæcicq; errores toto grassantur in orbe:
Ani·Dñi   pace beas longa, Vera et pietate Britannos: 1579
Iustitia moderans miti sapienter habenas.
Chara domi, celebrisq; foris, longæuaq; regnū
Hic teneas, regno tandem fruitura perenni.

5

6

Indicem huic operi tripartitum adiecimus. Primo, Alphabetico ordine singulos Comitatus. Secundo, nostræ distributionis ac tabularum seriem. Tertio, Iudicum itineraria, Initia, & dierum Iuridicorum ac Locorum ad Iudicia tam Ciuilia quam Criminalia exercenda constitutorum definitum tempus (Vulgus Circuitus ac assisas vocat) reperies.

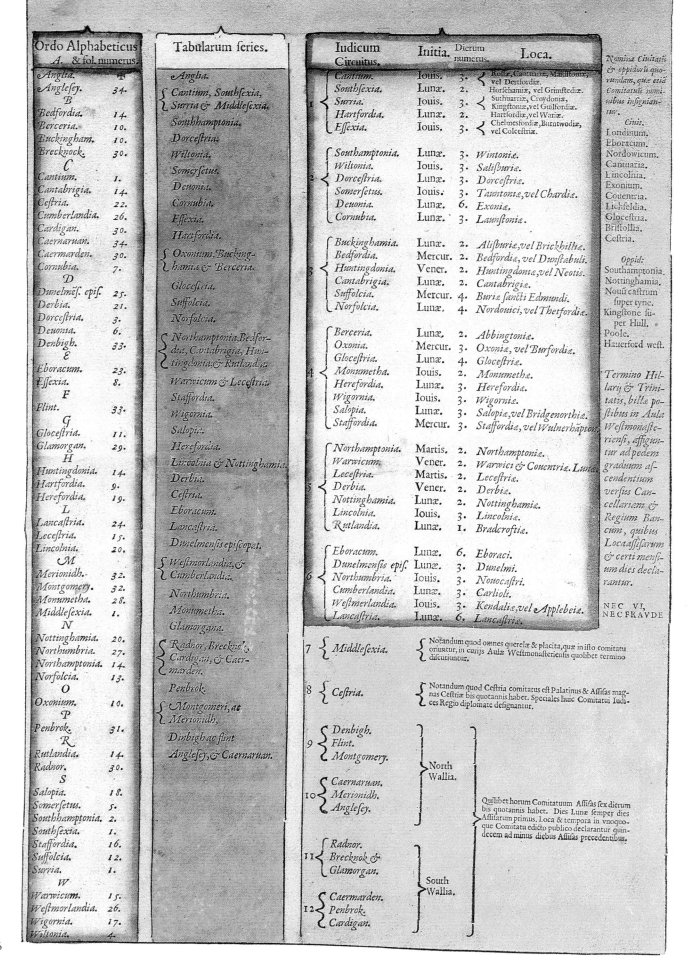

Ordo Alphabeticus A. & fol. numerus.

A	
Anglia.	1
Anglesey.	34
B	
Bedfordia.	14
Berceria.	10
Buckingham.	10
Brecknock.	30
C	
Cantium.	1
Cantabrigia.	14
Cestria.	22
Cumberlandia.	26
Cardigan.	30
Caernaruan.	34
Caermarden.	30
Cornubia.	7
D	
Dunelmeſſ. epiſ.	25
Derbia.	21
Dorcestria.	3
Deuonia.	6
Denbigh.	33
E	
Eboracum.	23
Essexia.	8
F	
Flint.	33
G	
Glocestria.	11
Glamorgan.	29
H	
Huntingdonia.	14
Hartfordia.	9
Herefordia.	19
L	
Lancastria.	24
Lecestria.	15
Lincolnia.	20
M	
Merionidh.	32
Montgomery.	32
Monumetha.	28
Middlesexia.	1
N	
Nottinghamia.	20
Northumbria.	27
Northamptonia.	14
Norfolcia.	13
O	
Oxonium.	10
P	
Penbrok.	31
R	
Rutlandia.	14
Radnor.	30
S	
Salopia.	18
Somersetus.	5
Southhamptonia.	2
Southsexia.	1
Staffordia.	16
Suffolcia.	12
Surria.	1
W	
Warwicum.	15
Westmorlandia.	26
Wigornia.	17
Wiltonia.	4

Tabularum series.

- Anglia.
- Cantium, Southsexia, Surria & Middlesexia.
- Southhamptonia.
- Dorcestria.
- Wiltonia.
- Somersetus.
- Deuonia.
- Cornubia.
- Essexia.
- Hartfordia.
- Oxonium, Buckinghamia & Berceria.
- Glocestria.
- Suffolcia.
- Norfolcia.
- Northamptonia, Bedfordia, Cantabrigia, Huntingdonia & Rutlandia.
- Warwicum & Lecestria.
- Staffordia.
- Wigornia.
- Salopia.
- Herefordia.
- Lincolnia & Nottinghamia.
- Derbia.
- Cestria.
- Eboracum.
- Lancastria.
- Dunelmensis episcopat.
- Westmorlandia, & Cumberlandia.
- Northumbria.
- Monumetha.
- Glamorgana.
- Radnor, Breckno', Cardigan, & Caermarden.
- Penbrok.
- Montgomeri, ac Merionidh.
- Dinbigh ac flint.
- Anglesey, & Caernaruan.

Iudicum Circuitus | Initia | Dierum numerus | Loca.

#	Iudicum Circuitus	Initia	Dierum numerus	Loca.
1	Cantium.	Iouis.	3.	Rossæ, Cantuariæ, Maidstoniæ, vel Dertfordiæ.
	Southsexia.	Lunæ.	2.	Horsehamiæ, vel Grimstediæ.
	Surria.	Iouis.	3.	Suthuarriæ, Croydoniæ, Kingstoniæ, vel Guilfordiæ.
	Hartfordia.	Lunæ.	2.	Hartfordiæ, vel Wariæ.
	Essexia.	Iouis.	3.	Chelmesfordiæ, Burntwodiæ, vel Colcestriæ.
2	Southamptonia.	Lunæ.	3.	Wintoniæ.
	Wiltonia.	Iouis.	3.	Salisburiæ.
	Dorcestria.	Lunæ.	3.	Dorcestriæ.
	Somersetus.	Iouis.	3.	Tauntoniæ, vel Chardiæ.
	Deuonia.	Lunæ.	6.	Exoniæ.
	Cornubia.	Lunæ.	3.	Launstoniæ.
3	Buckinghamia.	Lunæ.	2.	Alisburiæ, vel Brickhilliæ.
	Bedfordia.	Mercur.	2.	Bedfordiæ, vel Dunstabuli.
	Huntingdonia.	Vener.	2.	Huntingdoniæ, vel Neotis.
	Cantabrigia.	Lunæ.	2.	Cantabrigiæ.
	Suffolcia.	Mercur.	4.	Buriæ sancti Edmundi.
	Norfolcia.	Lunæ.	4.	Nordouici, vel Thetfordiæ.
4	Berceria.	Lunæ.	2.	Abbingtoniæ.
	Oxonia.	Mercur.	3.	Oxoniæ, vel Burfordiæ.
	Glocestria.	Lunæ.	4.	Glocestriæ.
	Monumetha.	Iouis.	2.	Monumethæ.
	Herefordia.	Lunæ.	3.	Herefordiæ.
	Wigornia.	Iouis.	3.	Wigorniæ.
	Salopia.	Lunæ.	3.	Salopiæ, vel Bridgenorthiæ.
	Staffordia.	Mercur.	3.	Staffordiæ, vel Wulnerhāptoniæ.
5	Northamptonia.	Martis.	2.	Northamptoniæ.
	Warwicum.	Vener.	2.	Warwici & Couentriæ. Lunæ.
	Lecestria.	Martis.	2.	Lecestriæ.
	Derbia.	Vener.	2.	Derbiæ.
	Nottinghamia.	Lunæ.	2.	Nottinghamiæ.
	Lincolnia.	Iouis.	3.	Lincolniæ.
	Rutlandia.	Lunæ.	1.	Bradcroftiæ.
6	Eboracum.	Lunæ.	6.	Eboraci.
	Dunelmensis epis.	Lunæ.	3.	Dunelmi.
	Northumbria.	Iouis.	3.	Nouocastri.
	Cumberlandia.	Lunæ.	3.	Carlioli.
	Westmerlandia.	Iouis.	3.	Kendaliæ, vel Applebeiæ.
	Lancastria.	Lunæ.	6.	Lancastriæ.

7 { Middlesexia. { Notandum quod omnes querelæ & placita, quæ in isto comitatu oriuntur, in curijs Aulæ Westmonasteriensis quolibet termino discutiuntur.

8 { Cestria. { Notandum quod Cestria comitatus est Palatinus & Assisas magnas Cestriæ bis quotannis habet. Speciales huic Comitatui Iudices Regio diplomate designantur.

9 { Denbigh. Flint. Montgomery. } North Wallia.

10 { Caernaruan. Merionidh. Anglesey. } Quilibet horum Comitatuum Assisas sex dierum bis quotannis habet. Dies Lunæ semper dies Assisarum primus. Loca & tempora in vnoquoque Comitatu edicto publico declarantur quindecem ad minus diebus Assisas precedentibus.

11 { Radnor. Brecknok & Glamorgan. } South Wallia.

12 { Caermarden. Penbrok. Cardigan. }

Right margin:

Nomina Ciuitatū & oppidorū quorundam, quæ etiā Comitatuū nominibus insigniuntur.

Ciuit.
Londinum.
Eboracum.
Nordowicum.
Cantuaria.
Lincolnia.
Exonium.
Couentria.
Lichfeldia.
Glocestria.
Bristollia.
Cestria.

Oppid:
Southamptonia.
Nottinghamia.
Nouū castrum super tyne.
Kingstone super Hull.
Poole.
Hauerford west.

Termino Hillarij & Trinitatis, billæ postibus in Aula Westmonasteriensi, affiguntur ad pedem graduum ascendentium versus Cancellariem & Regium Bancum, quibus Loca assisarum & certi mensum dies declarantur.

NEC VI, NEC FRAVDE

CONTENTS

INTRODUCTION

Christopher Saxton's Atlas: Background and Purpose

Christopher Saxton described himself in 1596 as 'of Dunningley in the parish of Westardesleye in the Countye of Yorke, Gent., of the age of fyftye twoo yeares or thereabouts'. Although he may have been a little uncertain of his own age there is no doubt of the esteem and admiration he acquired in the age in which he lived, a reputation neither tarnished nor eroded by the 400 years which have elapsed since he first produced his famous atlas of the counties of England and Wales in 1579. To the first Queen Elizabeth's subjects he was alongside individuals like William Shakespeare one of that select band who made blossom in England the summer flowers of the Renaissance which had known their springtime in fifteenth-century Italy. To such an extent was this the case that subsequently the intellectual thrust at home and expansion overseas of a comparatively small nation in this northern isle of Britain became increasingly identified with the main movements in the evolution of the modern world.

In Britain in the early part of of the sixteenth century many aspects of life trailed behind the Continent, but during the last quarter of that century the English made rapid progress in a number of intellectual and practical fields. One of these was cartography. Many of the principles, concepts and techniques of the subject can be traced back to Ptolemy's *Geographia*, a Greek manual on the construction and drawing of maps written at Alexandria about the year A.D. 160. The means by which the *Geographia* became known to Europe, however, was through its translation into Latin *c.*1406 and its subsequent widespread availability, first in numerous codices and then in print, some forty editions appearing before the year 1600. Outstanding among the basic principles enunciated by Ptolemy was the mapping of a landscape to a consistent reduced scale. The task of the Geographer, he declared, was to make a 'survey of the earth in its just proportions' and in support of that he continued 'the exact position of any particular place, and the positions of the various countries, how they are situated with regard to one another, how situated as regards the whole'. These words had far-reaching implications for such versatile minds as the Renaissance humanists who introduced a new age in which the reality of landscape came to be viewed and understood in geometric and mathematical terms. These concepts inspired by the *Geographia* were disseminated from Florence and Rome to the north of the Alps and thence to north-west Europe. The first signs of their arrival in England can be documented to the middle of the sixteenth century where they were to find fertile ground in view of the particular intellectual, religious, administrative and political upheavals which characterised Elizabeth's reign.

Until fairly recently it had remained something of a mystery why a seemingly obscure individual of yeoman stock hailing from the small Yorkshire hamlet of Dunningley lying between Leeds and Wakefield came to such prominence as a national map-maker. Although all the details in the story are not as fully known as one would wish the main strands can now be made explicit. The context points straight to the highest levels of government, namely the Crown and Elizabeth's ministers, the most prominent of whom was William Cecil, subsequently Lord Burghley, on whom posterity has conferred the accolade of being 'the most cartographically-minded statesman of his time'.

Burghley secured appointment to a sequence of high offices and was at or near the centre of the political stage for most of his long life. From as early as the 1550s to his death in 1598 he saw clearly the government's need for effective maps, and he used his immense influence to see to their production and preservation. The mapping of the whole realm on a suitable scale was a formidable undertaking, and much thought and careful preparation was clearly called for. The process can be detected in its initial stages soon after Elizabeth came to the throne. In 1561 John Rudd, a graduate of Clare College, a Fellow of St Johns, Cambridge and a prebendary of Durham Cathedral, who had already produced a 'platt' of England, announced that he intended to resume the work and to travel further 'for the setting forth thereof both fairer and more perfect and truer than it hath been hitherto'. Rudd expected to be away for two years and in order to assist him in this work no less a person than the Queen herself wrote to the Chapter at Durham ordering it to pay all the emoluments of his prebend during his absence.

John Rudd was one of those clerics who saw and experienced at first hand the full cycle of the religious changes in the sixteenth century but he was prepared to bend with the prevailing wind and this flexibility in matters of faith may well have been accommodated by the fact that his abiding interest lay elsewhere. As early as 1534 he had declared himself to have long been a student and teacher of the map-maker's art and some twenty years later he was presented to the Yorkshire vicarage of Dewsbury where he remained until 1570. Dewsbury is but three miles from Dunningley where it will be remembered Saxton lived. He would have been some ten or twelve years old when John Rudd arrived in Dewsbury and there is much to suggest that the two were associated over the ensuing years for in April 1570 Saxton was sent to Durham to collect £8 6s 8d 'for thuse of my master Master Rudde for his quarter stypend dewe at thannuntcyacon last'. The tenor of this statement and the nature of the mission points to the distinct probability that Saxton may well have progressed from being a pupil of John Rudd to becoming his assistant. Here must surely lie Christopher Saxton's initiation to the map-maker's art and his apprenticeship in the science of field surveying. John Rudd must undoubtedly have had a considerable formative influence on the much younger Saxton, but probably also deserves recognition for being a map-maker in his own right.

Nothing of John Rudd's cartographic work of this period in the 1560s is known to survive but a rival bid to map Britain was submitted to William Cecil by Laurence Nowell in 1563. His letter of application 'To . . . Sir William Cecil Knight, the Queens Majesties principal Secretarie' opens pointedly: 'I observe, most honoured Sir, that above all other monuments of the noble arts you take especial pleasure in geographical maps, and that you know how to make good use of them in your office to render increasing service of all kinds to the state'. Laurence Nowell was well acquainted with William Cecil at this time, being resident in his house as a tutor to the Earl of Oxford, one of Cecil's wards. Nowell knew that existing maps were not to Cecil's liking as was made so explicit in the same letter, ' . . . those who have hitherto undertaken to describe the country of England have not in all respects satisfied you; nor certainly (let me speak frankly) have they deserved even moderate praise. And this is not to be wondered at, because, without precise rule and without the support and judgement of any art, they have brought together in their maps certain imaginary locations and intervals of places either combining the reports of any others or by relying on the uncertain estimation of their own eyes'. Nowell's proposal to depict 'our country both as a whole and in all its parts, and also the several counties' does not, however, appear to have secured Cecil's approval even though Nowell was not without proven cartographic skill. From what subsequently transpired it would appear that Cecil was more inclined to favour the good field man in preference to the scholar and academic, in other words a survey at first hand in the landscape rather than a compilation of synthetic maps from previous sources. Evidence of his liking for this approach emerged in 1567 when both he and the government were considering how the newly pacified province of Ulster might best be garrisoned and secured. In May of the following year the Privy Council agreed to send a surveyor 'skilful in the description of countries by measure according to the rules of cosmography', having chosen the mature and experienced field man, Robert Lythe, to undertake the task.

In the early 1570s, what had been seen as an unfulfilled need in the cartography of Ireland would have applied equally to the main island of Britain. By this time, however, the government, and particularly William Cecil, had been provided with many opportunities to familiarise themselves with what exactly was needed, as well as the scale of the undertaking, the time required, the back-up support necessary and the nature and expertise of the individuals who would be needed if a successful outcome was to be ensured. It was a task which would have suited John Rudd, the mapping of large areas being his forte; he had already received the Queen's support for such work but he was now a man in his seventies and this was clearly a commission for a much younger man. The task ahead was a formidable one, requiring long periods in the field: stamina, health, concentration, adaptability and perseverance go most readily with the young. This was not all. There had to be experience, an eye for country, and a single-mindedness to see the work not only pursued according to a programme, but also finished on time. These were the likely qualities which John Rudd had detected and probably nurtured in his assistant Christopher Saxton during their field-surveying together; he was the kind of well-trained field man who would particularly suit William Cecil, for nothing but an original survey at first hand throughout the whole of the country could produce the end product now envisaged by the Queen's administrators.

There now seems little doubt that it was John Rudd who had the appropriate connections with royalty and with the upper échelons of Elizabeth's administration to recommend Christopher Saxton to continue with his kind of work and to gain for him an introduction to the appropriate officials at her court. The official who became most directly involved with Christopher Saxton was Thomas Seckford. He was a rich and public-spirited Suffolk gentleman who was Master of Requests to Queen Elizabeth and who in the last resort became Saxton's paymaster; such a procedure need cause little surprise as this was quite characteristic of the indirect method of government favoured by Elizabeth who, wherever possible, preferred to have the cost of public services borne by private subjects such as merchants, noblemen, courtiers and office-holders. The first mention which survives of this sponsorship, for that is in reality what it amounts to, is dated 11 March 1576, when the Privy Council issued the following order: 'A placart to . . . Saxton servant to Mr Sackeforde. Mr of the Requests to be assisted in all places where he will come for the view of mete places to describe certen counties in Cartes being thereunto appointed by her Ma[jes]tes bill under her signet.' This was signed by William Cecil but now as 'Lord Burghley, Lord Treasurer' and ten other councillors.

The actual field-work and the production of the maps must pre-date the signing of the 'placart' by some years, as on 11 March 1574 Saxton received a grant of land 'for certain good causes grand charges and expenses lately had and sustained in the survey of divers parts of England . . . '. Furthermore, proof copies of the maps after having been engraved were being sent to Lord Burghley as early as 1574 so that there is much evidence to support the dating of the start of the survey in the early 1570s. Saxton would then have been in his late twenties and presumably would have had access to any previous work; therein lies a mystery as to what happened to John Rudd's earlier endeavours. How much, if any, was incorporated into Saxton's maps is at present an unanswerable question but if the new specification could have been met from what existed it is likely that the north-country maps would have appeared first. They did not and indeed a quite different scheme is apparent. If it is reasonable to assume that the dates borne by the engraved maps give an approximate indication of the sequence in which the counties were surveyed, then Norfolk comes first followed by Oxford, something of an exception, since there is a clear indication of a determined effort to complete the coastal counties of south-east and southern England before the Midlands and then the northern counties. All of England was completed by 1577, and the survey of Wales followed in 1578.

The order in which the counties were surveyed and the fact that proof copies of the maps when engraved were sent directly to Lord Burghley is further confirmation that this mapping exercise was an 'official' survey promoted by the Crown on the advice of the Queen's ministers, as an act of deliberate policy designed to produce maps for the purposes of national administration and, equally important, those of the defence of the realm. This can be seen from the use Lord Burghley made of the proof sheets sent to him. They were bound together into a veritable *atlas factice* with the individual maps being annotated in his own hand with corrected spellings of place-names, lists of justices and gentry, notes on coastal defence, landing places, local military organisation and on inserted leaves tables of roads and posts. These manuscript additions were being made until about 1590 but the greatest concentration of them, not unexpectedly, relates to the period just before the arrival of the Spanish Armada off the shores of southern England in 1588.

Christopher Saxton's Atlas and the Chatsworth Exemplar
On 20 July 1577 Queen Elizabeth granted Christopher Saxton a licence for the exclusive publication of his maps for a period of ten

years. In what form and on what scale Saxton first drew his maps and whether he submitted such drafts to his employers for approval are questions which cannot be answered for lack of surviving evidence. Once, however, the final manuscript drafts were made for the engravers there is distinct evidence of the process of compilation whereby each map had to be adjusted to a scale to fit into a rectangular frame within the space provided by a single copper-plate, the image to be printed subsequently on a royal sheet of paper about 25 by 20 inches in size. With the exception of the large county of Yorkshire which required printing on separate leaves subsequently to be butted and folded, all the other counties were adjusted to a roughly similar size and regular format to be readily bound into a book to form an atlas. No copy bears the title 'atlas' and indeed the use of this term does not appear until 1595 when it was used by the eminent Flemish cartographer, Gerhard Mercator, on the title page of his world atlas. Mercator was not the first to attempt the printing of a collection of maps to a consistent format in order to be readily bound together, a model had been provided as early as 1570 by the *Theatrum Orbis Terrarum* of Abraham Ortelius. There is no doubt that this specific model did much to focus aspiration in Elizabethan England and, according to a contemporary writer, provided the prototype for the assembling of Saxton's individual maps into a bound format; in this respect, once again, one must view the atlas as an English development of an activity pioneered in Renaissance Europe. The atlas of 34 county maps, together with 'Anglia', a small-scale map embracing England and Wales, and a number of preliminaries all bound into one folio volume was entered on behalf of Christopher Saxton at Stationers' Hall in 1579. This actual event was anticipated in 1577 by the telling comment in Holinshed's *Chronicles* that 'my countrymen eare long shall see all Englande set foorth in severall shyres after the maner that Ortelius hath dealt wyth other countries of the mayne, to the great benfite of our nation and everlasting fame' of those who were responsible for its production. Ortelius's achievement had clearly awakened a patriotic spirit of emulation in England.

The actual number of extant copies of Saxton's atlas is, for a variety of reasons, not precisely known but those available for study in libraries are such that it is possible to place the Chatsworth copy appropriately among them. Previous carto-bibliographic studies of surviving copies reveal that the atlas in its early form, that is up to the end of Elizabeth's reign and before the more extensive changes wrought in the 1640s, display differences in make-up and individual sheets which enable them to be arranged in an approximate chronological sequence. It is now clear that the preliminary leaves and the 34 county maps engraved between 1574 and 1578 were not printed in one batch and bound to provide what may be described as a standard edition. This is borne out by the baffling variety of watermarks in the papers used. They are all of the bunch of grapes type but much differentiated by the addition of letters and emblems which points to manufacture in France before c.1600. What seems to have happened is that the maps, which were also available as single copies, were printed and bound on demand, thus permitting interim alterations to be made to the copper-plates and letter-press, these then providing the sheets for the atlas volumes. Any particular atlas may therefore be assigned its place in the sequence by collating it with other known copies and by relating it to the known alterations and additions which appeared in the atlases during their printing history. A number of diagnostic criteria may therefore be applied to the Chatsworth atlas in the following way.

Most of the original copies have the engraved frontispiece portraying Queen Elizabeth enthroned under a canopy as the patron of Geography and Astronomy which are represented respectively by bearded male figures, the one on the right bearing compasses and a globe, the other an armillary sphere; other figures of a smaller size are Righteousness and Peace kissing each other in a medallion above the Queen's head and below her feet the two aspects of map-making, the topographic and the geodetic, represented respectively by surveyors, one drawing a map and another shooting the stars to fix his spherical position. The cartouche in the lower quarter, inscribed with a Latin verse of six lines in praise of Elizabeth's benign reign, bears the date 1579. This detailed engraving, though unsigned, has been attributed to Remigius Hogenberg rather than the other possible claimant, Augustine Ryther, and appears in two states. In state I the Queen's dress has elaborate jewelled ornamentation and her robes lie in a hard horizontal line across her knees. In state II the Queen's forearms and dress have been re-engraved, her jewellery much simplified and more natural folds of the drapery replace the previous arrangement. Of the two states the former one is extremely rare and this has prompted the suggestion that at an early stage, probably even in 1579, the Queen was less than happy with it and ordered the changes to be made. In this respect the presence of either the one or the other is not particularly helpful for dating purposes. In many copies, however, this attractive frontispiece has been removed, and lost; fortunately in the Chatsworth exemplar it has survived, richly coloured and in a good state of preservation in the later state II.

The Index of Maps, the only page of the atlases to be printed on movable type, is found in no less than four different type-settings, usually designated with the letters A, B, C and D. Setting D is the most commonly found of the settings and this is the one in the Chatsworth copy. The heading takes up four lines of print followed by three columns in which the listings are much fuller than in the other settings. The first column provides a list of the counties in alphabetical order with number reference 1–34. 'Anglia', which appears first, has no number but is designated with a Maltese cross. The next column to the right lists again the 35 maps but here arranged according to the order in which they appear in the atlas starting with 'Anglia' and ending with 'Anglesey & Cearnauan'. The third is a table of judge's circuits and assizes. This fuller and more analytical treatment clearly differentiates setting D from the others and, besides being the latest, becomes the definitive arrangement; this helps to date the Chatsworth atlas to c.1590 with a good degree of confidence.

The reason for the above assuredness rests on the fact that setting D is in the majority of cases accompanied by a double-page spread exhibiting coats-of-arms and a statistical table of the counties. Both have been engraved on a single plate, the left-hand page illustrating 84 coats-of-arms the last one of which in the bottom right-hand corner is blank. The first 65 are coats-of-arms of peers while the remainder are those of officers of state. Among the latter are Sir Christopher Hatton, No.66, who is noted as Chancellor, an office to which he was appointed in 1587 and held until his death in 1591. Likewise, Sir Thomas Heneage, No.69, is designated Vice-C[hamberlain] an appointment he secured in 1589. The dates of these tenures of office indicate clearly that this plate could not have been engraved before 1589 and, probably, not after 1591. The right-hand leaf consists of a formidable statistical table listing in the first column the 52 counties of England and Wales, and then in rows numbers of such items as their cities, bishoprics, market towns, castles, parish churches, rivers, bridges, chases, forests and parks. At the foot of the double-page spread is a small 'Scala Milliarium' giving the relative lengths of the long, middle and short miles.

It is profitable to consider the first map in the atlas, 'Anglia', with the preliminaries as it, by existing in two states, throws further light on the question of dating. In state I the borders consist of a decorated outer and a plain inner section, whereas in state II the latter section is graduated for longitude and latitude. Although the meridians and parallels are not actually drawn across the face of the map, if the corresponding pairs of graduations are lined up with each other they will be seen to form the Donis projection, known also by the name of 'trapezoid and trapeziform', since this is the form the graticule assumes. The term 'Donis' is one of the forenames of Nicolaus Germanus, a Benedictine monk who from c.1460 dedicated himself to the promulgation and improvement of Ptolemy's *Geographia* and its associated maps. When, eventually, the *Geographia* appeared in its printed as distinct from its manuscript form, almost all the editions made use of the trapezoidal projection. It enjoyed such popularity on the Continent subsequently that it is not surprising to find it being used in Britain in the sixteenth century. The graduations on 'Anglia' are for whole degrees, ten and two minute intervals. The placing of these after the map image had been engraved on the copper-plate revealed that a slight shift was required in their alignment. They were not made to coincide exactly with the rectangular neat lines, accuracy seemed to over-ride aesthetic considerations. Such sophistication relates 'Anglia' to Saxton's other great masterpiece, the large general wall map, *BRITANNIA INSVLARVM IN OCEANO MAXIMA . . .*, which was published in 1583; more about this rare and outstanding map will be seen below but for the present it gives an added piece of evidence for the Chatsworth exemplar being late in the sequence of atlases. 'Anglia' in its own right is a particularly attractive map; the Latin title is within a one-sided large strap-work cartouche in the top right-hand corner surmounted by the Royal Arms with supporters, below it are the arms of Thomas Seckford. The top left-hand corner is also filled by a one-sided strap-work cartouche surrounding the 'Index omnivm comitatvvm' which lists the counties with reference numbers; these facilitate the identification of the counties which are not named on the actual map. To balance the whole there is in the south-west corner a richly decorated linear scale with single-handed dividers over it incorporating Saxton's name as author and Augustine Ryther's as engraver.

The maps of the counties in Saxton's atlas provided a new standard of cartographic portrayal in Britain. There were in all 34 maps of counties; nine were combinations of adjacent counties sometimes in twos, but up to five in the group around Cambridgeshire. No credible explanation for this arrangement has ever been promulgated nor is there any clue why the twenty-four which appear as single counties were so distinguished. All the maps, in needing to be adjusted to the size of the copper-plate and to the sheets of paper show quite a variation in scale; an approximate representative fraction for the smallest, Lancashire, is 1:301,000 and for the largest, Monmouthshire, 1:141,000. The constraints imposed by paper size and book format thus involved an immense labour in reduction and compilation, but this should not conceal the fact that in the field Saxton would have used a common unit of linear measurement.

The topographical features delineated on the county maps consist of elements of both the physical and cultural landscapes. Upland relief is in the form described as 'mole-hills' or 'sugar-loaves', usually hatched on their south-eastern sides to give the impression of shadows cast by a setting sun. The map of Oxfordshire and its surrounding counties is an intriguing exception, the hatching being inscribed on the western sides. The hills vary in size and are intended to convey an impression of high land only rather than any attempt being made to show actual magnitude: this becomes clear when other topographic features demand a presence, the space-greedy mole-hills have to surrender pride of place. Rivers are given considerable prominence, and though in places they appear with exaggerated widths somewhat stylised in their upper reaches rather than dendritic, it does seem as if Saxton, quite sensibly, viewed the landscape in terms of valley systems. Lakes and the sea are shown as stippled areas, except in the case of the latter where the space is occupied by cartouches and other embellishments.

Saxton's representation of the cultural landscape concentrates on recording a hierarchy of settlements almost all of which are named. Although no key is provided, normally there is a six-fold classification of places apparent, beginning with a small circle with a dot in its centre. The most prominent symbol superimposed on these is that for cities, depicted by a group of churches and houses, a cathedral prominently differentiated by a large cross; towns are shown by a symbol embodying three-spired churches; parish churches by a single-spired church symbol; chapelries or hamlets by a gabled house; castles and manor houses have their own symbols, the former resembling an edifice with two flanking towers while parks, whether associated with larger establishments or not, are realistically depicted as areas surrounded by a ring of palings. Not all recorded settlements fit into this tidy framework; there are, for example, some small circles, often un-named, which may represent isolated or single farmhouses. The settlement geography across county boundaries where two county maps overlap sometimes shows inconsistencies; named places appear on each sheet but with different symbols. The explanation for such anomalies may reflect lapses in Saxton's memory as he moved from one county to another or when he drafted the maps at different times. It is more likely, however, that they reflect the mis-interpretation of Saxton's instructions and drafts by the engravers of whom there were several and who were not working in the same place and so lacked Saxton's direct supervision — he being in the field distant from the engravers' workshops of London.

There is little evidence to support the contention that Saxton himself engraved any of the maps for his atlas and so, when one admires their distinctive elegance and charm, a tribute is owed to the skilful hands that performed this essential part of the publication process. Of those who engraved their names on their maps four were Flemings who had emigrated to Britain; Remigius Hogenberg signed nine maps, Lenaert Terwoort five, Cornelius de Hooghe and Johannes Rutlinger one each. Of the 13 unsigned maps, five, to judge from their style, may also be the work of Flemings. On the other hand, though initially under Flemish tutelage, the making of Saxton's atlas is now recognised as a crucible for English map-engraving. Augustine Ryther, proud to be registered an Englishman by the addition of *Anglus* after his name, engraved five maps, Francis Scatter two and Nicholas Reynolds one, but so effectively did these craftsmen work in harmony on this project that the term 'Flemish-English school' may justly be used to characterise their artistic efforts. On so many of the county maps it is the elaborate decoration which may at first concentrate one's gaze or, as some may claim, distract one's attention from the maps themselves. Neptune with his trident embracing a maiden off the coast of Anglesey symbolising the union of land and sea, galleons sailing or engaged in combat, fishing vessels, grotesque sea monsters disturbing the surface waters enliven the sea voids offshore the maritime counties, all these give pleasure to the eye and make these maps such attractive works of fact and fancy.

Even more spectacular are the numerous cartouches in a style markedly Flemish which had its origins in the fusion of two quite separate elements, both emanating from Italy. This particular style has become known as strapwork, so called because it resembles snipped, rolled and curled up edges of leather saddlery and shields. It was not allowed thus to stand alone but became mixed with elements of the grotesque, a mode of decoration copied by the Renaissance Italians from the designs they discovered on the walls of excavated ancient Roman houses. The new school of Flemish decoration took the basic strapwork motifs and elaborated them hugely, treating them in fancy as solid structures of wood or even of stone, piercing them with apertures, extending them with branches and strengthening them with cross-pieces. Even all that did not satisfy as these structures were animated with men, women and babes perched and clambering about the strapwork as well as masks, birds, animals, clusters of fruit and ribbons. England in turn fell victim to this cartouche mania such that this oddly-contrived feature of sixteenth-century Renaissance design gave much of its character to Tudor engraving and appears to have blended well with the exuberance so characteristic of the masques and pageantry of Elizabeth's court.

Saxton's maps in the plain form in which they left the engravers' hands are rare indeed. The use of colour on the maps had by Saxton's day assumed standard conventions. Rivers, lakes and water are 'washed', the contemporary technical term, in blue, woods are green, hills stand out in brown and green, settlements in red and county boundaries are outlined in contrasting shades. Colour is fairly consistently applied through a set of maps bound in one contemporary copy such as that at Chatsworth but since individual copies are the works of different colourists whose choice of colours, outside the conventional ones, is idiosyncratic, no atlas can therefore be exactly like another; herein lies that essential individuality which makes a reprint of the Chatsworth exemplar such a worthwhile endeavour.

The library at Chatsworth is one of the largest in England still in private hands and has been amassed over several generations by acquisitions and purchases. The provenance of this particular copy of Saxton's atlas is difficult to unravel since it was rebound in the early part of the nineteenth century and the original end papers which may have provided clues to previous ownership have been for ever lost. The back of the page on which the coats-of-arms are engraved is stamped with the name of Henry Cavendish, the renowned scientist, who amassed a very fine library which passed in due course to the sixth Duke of Devonshire; he moved it to Chatsworth on the completion of alterations to the Long Gallery to house his books. The library was re-catalogued in 1879 but the description did not directly associate this set of maps with Christopher Saxton with the result that the real significance of this atlas was not evident until recently.

Christopher Saxton's Surveying Techniques
Christopher Saxton has left to posterity little but silence about the surveying techniques he used to achieve his highly successful map-making. Over the last sixty years several scholars have attempted to provide answers, their verdicts oscillating between the two major field procedures, triangulation and traversing. In the 1920s the views held came down firmly in favour of triangulation but in the late 1930s the pendulum swung towards traversing owing mainly to a confusion in the precise use of terms. A collective uncertainty and vagueness has ensued until fairly recently when it has been suggested that progress could be made with this problem if the two great masterpieces, the atlas containing the small general map, *Anglia,* with the thirty-four county maps and *BRITANNIA INSVLARVM IN OCEANO MAXIMA . . . ,* the large general map of 1583, be examined as the outcome of a single integrated mapping operation. Furthermore, the undoubted quality and sophistication of the two general maps have not been sufficiently explored and explained, either in terms of the

field survey which was necessary for their compilation or in the high measure of control which they exhibit. Although no specific and personal information may survive, generations of surveyors have faced the same challenge — a vast landscape to be put on paper. A more field-oriented approach is therefore needed in conjunction with an appreciation of the contemporary state of national and international affairs.

In order for the general maps to have been drafted so successfully, the field techniques required a control which is not provided by open traversing along roads using polar co-ordinates unless their terminal points are fixed by other means. Moreover, had all this information about roads been collected in the course of Saxton's survey, there is little doubt that he would have placed them on the maps, such was their economic, social, and strategic importance. Even the Fosse Way, requiring but a straight line across several county maps, has not been inserted, while Lord Burghley had to go to much trouble to add in manuscript lists of the 'ways, distances and postes' in his *atlas factice*. The general maps reveal further advanced features. The use of the Donis projection, the placing of the graticule, the choice of the central meridian and its relationship to a prime meridian passing through the Azores — all point to continental antecedents. The most likely agent for the diffusion to Britain of such sophisticated practices was John Dee. After Robert Recorde's death in 1558, John Dee became the most influential teacher and adviser on scientific subjects in England, and he retained this position for the next quarter of a century. He had formed close relationships with the two leading continental cartographers, Gerhard Mercator and Gemma Frisius. The influence of the former can be detected in the projection; the genius of the latter is seen in his *Libellus de locorum describendorum ratione*, first published in 1533, a copy of which graced the shelves of John Dee's library at Mortlake.

It was Gemma Frisius who appreciated that the astrolabe, which acquired its position by gravity in the vertical plane as an altitude-measuring device, could also function similarly in azimuth when levelled horizontally. The *Libellus* goes on to explain and illustrate how a third point in the landscape can be fixed from two known points by intersecting rays. The other requirement for this early form of triangulation was the known length of one side of a triangle. This was obtained either by direct measurement or by the use of a system of right-angled similar triangles. The *Libellus*, therefore, offers the first complete treatise on triangulation, and is not only a monument to the original mind of Gemma Frisius but also an important landmark in the history of cartography. John Dee would have become familiar at first hand with this new technique during his residence at Louvain during the period 1546–1548.

In addition to techniques, successful triangulation required appropriate and good angle-measuring instruments. These were also to hand in Britain, and were again based on antecedents and skills of continental origin. In the sixteenth century Flemish engravers on metal came to London to escape religious persecution, and so provided the nucleus of a scientific instrument-making trade. Thomas Gemini (fl. 1524–1562) was a leader in these skills, and he was followed by Humfrey Cole (1520–1591), the first of our English mathematical-instrument makers. His surviving instruments, among them theodolites, are superb pieces of engraving and dividing on fine metalwork. He must certainly be regarded as the leading instrument maker of the Elizabethan age, and one, moreover, known to and associated with William Cecil, Lord Burghley, who through Thomas Seckford, would not have neglected to equip Saxton with the best available instruments.

The involvement of the Crown and Government in map-making, has been explored, and due emphasis has been given to the organisational and administrative purposes which map-making was destined to serve. In so doing, possibly too much has been made of this issue to the detriment of one equally telling. For example, it has been claimed that the consistent appearance of those paled parks on Saxton's county maps is to do with the pride and prestige of the emerging Elizabethan gentry. No doubt there is much truth in that. Equally, if not more significant, may be the fact that from 1536 the Government was deeply concerned that 'the brede of horses is sore decayed', and all the owners of parks of up to two miles in circumference had to keep two breeding mares and four mares in parks over that size. This statutory obligation was still in force everywhere in the realm in 1560, a year when a return was demanded. This reflected the growing tension in relationships between Spain and England. In this same context, has sufficient emphasis been placed on the question of whether Saxton's mapping was an essential part of the military preparedness? Elizabeth's reign was one requiring continuous vigilance, and in the late 1560s the situation at home and abroad was becoming more threatening and dangerous. In subsequent history such times have frequently been those most conducive to investment in and production of maps. Speed was of the essence. The techniques, the instruments, the input from the *cognoscenti* at home and abroad, government support and indirect finance, the right man for the job — all these set a perfect stage, but for one omission. How to do the actual mapping consistently county after county. In five or so summer seasons? Less than one month for each county? A vital feature of the Elizabethan context has been overlooked. Who could have been relied upon to provide Saxton, a stranger, with the landscape data required in every county?

In 1567 the build-up of a strong Spanish army in the Netherlands constituted a real threat, and was so understood by Elizabeth and her Privy Council. As the political situation deteriorated the musters, which had formerly been called every three years, were now more frequent. On 26 March 1569 the Privy Council issued a directive to all counties to hold a general muster of all able-bodied men over the age of sixteen. Christopher Saxton would have been an active twenty-five or twenty-seven-year-old. With the musters went the beacon system. This constituted a nationwide, well-organised, intervisibly-linked communication system, which was centuries old by 1569: orders about it appear in the State Papers at intervals from 1324; these orders become more numerous in Elizabeth's reign. They reveal a comprehensive system being brought to a high state of readiness in the early 1570s and one regarded as an important part of the defence of the realm. Those responsible gave the system great thought so as to bring it to perfection. There was, for example, a chain of command from the Crown to county level, through Lords Lieutenant to Deputy Lieutenants to Justices of the Peace to High Constables of the Hundreds and Petty Constables of the Parishes. Christopher Saxton, therefore, need not have arrived cartographically empty handed in an unfamiliar landscape of a county. There was a maintained, manned set of known intervisible viewpoints, the system incorporated overlapping connections between the counties.

The word beacon, of Teutonic origin, is first introduced into the Latin of an ordinance in 1372, where it appears in the form 'common signals by fire *signum per ignem* on hills and high places called Beknes'. The repetition of this same phrase in a letter of the Privy Council to the Justices of the Peace and other officers in Wales on 10 July 1576 is particularly noteworthy.

> An open Lettre to all Justices of peace mayours
> & others etc within the severall Shieres of Wales.
> That where the bearer hereof Christofer Saxton is
> appointed by her Maiestie vnder her signe and
> signet to set forth and describe Coates[Cartes] in parti-
> culerlie all the shieres in Wales. That the said
> Justices shalbe aiding and assisting vnto him to
> see him conducted vnto any towre Castle highe
> place or hill to view that countrey, and that he may

be accompanied with ij or iij honest men such as
do best know the cuntrey for the better accomplish-
ment of that service, and that at his departure
from any towne or place that he hath taken the
view of the said towne do set forth a horseman
that can speke both welshe and englishe to safe
conduct him to the next market Towne, etc.

The message being relayed in this letter is of vital importance, in that it provides the only direct evidence so far found which gives unequivocal insight into the methods of survey used by, and the local support available to, Saxton. It is reasonable to assume that what applied to Wales would be equally applicable to the counties of England. The specific mention of 'towre Castle highe place or hill' is consistent with the use of triangulation, and the Justices of the Peace are instructed not only to see Saxton conducted to such high viewpoints but also to ensure that he was accompanied by two or three honest men who knew the surrounding landscape. Although the word 'beacon' does not specifically occur in this letter, the whole tenor of the message is entirely consistent with the known organisation of the beacon system. In the first place it was under the jurisdiction of the Justices of the Peace and secondly the watch was normally to be kept by two or three trustworthy men. The other important point made in the 'open Lettre', the provision of men who 'best know the cuntrey', can also be accommodated in the contemporary orders for the beacons. Those who watched, besides being 'wise and vigilant' had to be men of 'understanding' who were able to recognise precisely their nearest-neighbour beacons, that is 'the watchers on the hills both on the coast and inland, taking heed not to be deceived by other fires'; and again, 'the watchers must take hede that they fier not ther beakons unadvisedlie uppon any other fyers whatsoever shall fortune, in any place then uppon the view'. A beacon alarm had to be differentiated from other bonfires. There had been genuine errors of identification as well as cases of vandalism, hoaxes and the activities of *agents provocateurs*. By 1586 such incidents brought a swift and stern reaction from the Privy Council.

What is made clear by these references is that the watchers at a particular beacon would have been able to direct Saxton's theodolite open sights to a round of known places and to the neighbouring beacons, the system thus providing a logic to a landscape which would otherwise be strange, bewildering and unknown. Furthermore, he could rely on the system being in operation and available to him everywhere in the landscape in the summer months. This brings out the absolutely vital importance of his being in possession of the 'placart' or pass issued to an individual by the government or one of its agents. Such a pass was issued to Saxton on 11 March 1576. The problems arising from not having one, or, having had one, not having had it subsequently ratified, are conspicuously exemplified in the case of the other notable Elizabethan surveyor, John Norden, the non-renewal of whose pass led to the abandonment of his proposed country-wide re-survey, his *Speculum Britanniae*.

A large number of Elizabethan beacons have been documented, enough indeed to lend support to the view that they could have formed the instrument stations for a triangulation survey as in fact many of them did 200 years later for the Ordnance Survey. Unlike that later government institution, however, Saxton would have developed his triangulation on several base-lines and it is almost certain that the field-work was carried out on a county schedule. The modern principle of making a map by working from the whole to the part does not fit the Elizabethan surveying scene, and independent surveys of individual counties subsequently joined together would more likely have been the scheme adopted. His arrangement of triangles would also not necessarily have resembled the ordered network of well-conditioned triangles with which surveyors nowadays are familiar. The preliminary reconnaissances required to bring such order and form are difficult to accommodate within the time which Saxton actually took to complete the work. Much more feasible would appear to be the use of a resection technique, whereby the plotting of a station occupied can be effected. The field data required are: two visible stations their positions already established, and a ray from one of them to the station occupied. This, plus the more usual establishment of stations by intersection of rays, would have allowed Saxton to progress through the landscape. One must envisage him taking a comprehensive round of rays from a particular station to all points of topographic detail, together with the beacons within view and all these angles booked relative to a reference object. An equally comprehensive plotting of the data from his survey books during the progress of the field work would have ensured a smooth and orderly sequence to the project.

Christopher Saxton's cartography was greatly esteemed by his learned contemporaries as well as by succeeding generations: William Camden (1551–1623) dubbed him 'optimus chorographus' while printings from his original copper-plates remained a commercial proposition for the next 150 years in addition to the extensive plagiarisation of his maps by others. Although initially the mapping was undertaken to meet the needs of administrators who found it necessary to bring spatial dimensions into the deliberations of Government, it was also an aspect of the humanist Renaissance awakening in English life. In this broader social context, the English could for the first time take effective visual and conceptual possession of the physical landscape in which they lived. For the first time, people could perceive space and place in considerable detail; their country, their county, their home, and, pointing to a spot on a coloured 'paper landscape', they could feel as well as declare, 'Here I belong'. Such psychological security was as dramatic as it was deep, reinforcing a profound sense of regional and national identity. Fostered also by Christopher Saxton's atlas was the new emotion of experiencing a sheer delight in maps as objects of beauty as well. The decorative and eye-catching qualities of the maps assisted the wide circulation of the atlas, particularly among the country's educated élite for these cultural as well as utilitarian reasons. Such use over the centuries has led to the inevitable destruction and loss of many copies and most of those that do survive are the cherished possessions of Libraries and other learned institutions.

The aim and true purpose of reprinting such rare old maps is to make them more widely and readily available for many more people to enjoy in addition to the already committed map-lovers. It is with those thoughts in mind that we go to press in admiration and memory of 'the father of English cartography', Christopher Saxton.

William Ravenhill
St Nectan's Day 1992

SUGGESTED FURTHER READING

Peter Barber, 'A Tudor Mystery: Laurence Nowell's Map of England and Ireland', *The Map Collector,* 22 (1983), pp.16–21.

G.R. Crone, *Maps and their makers* (London, 1953), p.108.

Ifor M. Evans and Heather Lawrence, *Christopher Saxton Elizabethan Map-maker* (Wakefield, 1979), p.44.

Sir George Fordham, 'Christopher Saxton of Dunningley His life and work', Thoresby Society's *Miscellanea,* 28 (Leeds, 1928), pp.356–84 and 491.

Gemma Frisius, *Libellus de locorum describendorum ratione* annexed to Petrus Apianus, *Cosmographia Liber* (Antwerp, 1533).

R.T. Gunther, 'The Great Astrolabe and other Scientific Instruments of Humfrey Cole', *Archaeologia,* 76 (1927), pp.273–317 and 'The Astrolabe' its uses and derivatives', *Scottish Geog. Magazine,* 43 (1927), pp.135–47.

J.B. Harley, 'Christopher Saxton and the first Atlas of England and

Wales, 1579–1979', *The Map Collector,* 8 (1979), pp.2–11 and 'The Map Collection of William Cecil, First Baron Burghley 1520–98', *The Map Collector,* 3 (1978), pp.12–19.

Edward Heawood, 'Some Early County Maps', *Geographical Journal,* 68, No.4 (1926), p.325.

Gordon Manley, 'Saxton's Survey of Northern England', *Geographical Journal,* 83 (1934), pp.308–16.

David Marcombe, 'Saxton's Apprenticeship: John Rudd, a Yorkshire Cartographer', *Yorkshire Archaeological Journal,* 50 (1978), pp.171–5.

A. Pogo, 'Gemma Phrysius, his method of determining differences of longitude by transporting timepieces (1530), and his treatise on triangulation (1533)', *Isis,* 22 (1934–5), pp.469–485.

R.A. Skelton, *Saxton's Survey of England and Wales* Imago Mundi Supplement, No.VI (Amsterdam, 1974) and *County Atlases of the British Isles 1579–1703* (Carta Press, London, 1983), pp.112–119.

Sarah Tyacke and John Huddy, *Christopher Saxton and Tudor map-making,* British Library (London, 1980).

William Ravenhill, *John Norden's Manuscript Maps of Cornwall and its Nine Hundreds* (Exeter, 1972), pp.26–7.

William Ravenhill, 'Projections for the large general maps of Britain', *Imago Mundi,* 33 (1981), pp.21–32 and 'Christopher Saxton's surveying: an enigma', in *English Map-Making 1500–1650,* edited by Sarah Tyacke, The British Library, (London, 1983), pp.112–119.

E.G.R. Taylor, 'The Earliest Account of Triangulation', *Scottish Geographical Magazine,* 43 (1927), pp.341–5.

28

CATALOGVS *Vrbiū, Episcō; oppido Merciū, Castrō, Eccle, parochialiū, Fluui illustriū, Pontiū, Lucorū, Saltorū, Septorūq̄ omniū, quæ per totam Angliā Walliāq̄, in Vnoquoq̄ comitatū continentur, quemadmodū suis locis in Chorographicis Angliæ Walliæq̄ tabulis (vbi sui cuiq̄ nomē adijciūt) illustrissime referūtur. Numerus verò eorum omnium quæ in hāc serie colliguntur, ad imum huius indicis assignatur, sicuti infra videre licet.*

Comitatus	Vrbiū	Episc	OpM	Castr	EcPa	Fluū	Pon	Lucō	Salto	Septo
Cantium	002	002	017	008	398	006	014	000	000	023
Southsexia	001	001	018	001	312	002	010	000	004	033
Surria	000	000	006	000	140	001	007	000	004½	017
Middlesexia	002	001	003	000	073	001	003	001	000	004
Southamptonia	001	001	018	005	248	004	031	000	009	029
Dorcestria	000	000	018	006	248	004	029	001	002	012
Wiltonia	001	001	021	001	304	005	031	001	004	022
Somersetus	003	002	029	001	385	009	045	000	002	018
Deuonia	001	001	040	003	394	023	106	000	000	023
Cornubia	000	000	023	006	161	007	031	000	000	009
Essexia	001	000	021	001	415	007	028	000	001	046
Hartfordia	000	000	018	000	120	001	024	000	000	023
Oxonium	001	001	010	000	208	003	026	000	001	009
Buckinghamia	000	000	011	000	185	002	014	000	000	015
Berceria	000	000	011	001	140	003	007	000	004½	013
Gloecstria	001	001	020	001	280	012	022	001	002	019
Suffolcia	000	000	028	001	464	002	032	000	000	027
Norfolcia	001	001	026	000	625	003	015	000	000	000
Rutlandia	000	000	002	000	047	000	001	000	000	004
Northaptonia	001	001	011	002	326	005	024	000	003	023
Huntingdonia	000	000	005	000	078	001	005	000	000	007
Bedfordia	000	000	010	000	116	001	006	000	000	012
Cantabrigia	000	000	006	000	163	001	007	000	000	005
Warwic	001	002½	012	001	158	007	021	001	000	016
Lecestria	000	000	011	002	200	001	010	000	002	013
Staffordia	001	001¼	012	005	130	013	019	001	001	038
Wigornia	001	001	007	003	152	005	013	001	002	016
Salopia	000	000	013	013	170	018	013	000	007	027
Herefordia	001	001	008	007	176	013	011	001	002	008
Lincolnia	001	001	026	002	630	009	015	000	000	013
Nottinghamia	000	000	011	000	168	005	017	000	001	018
Darbia	000	000	008	004	106	013	021	000	001	034
Cestria	001	001	009	003	068	009	019	000	002	018
Eboracum	001	001	046	014	563	036	062	004	008	072
Lancastria	000	000	018	006	036	033	024	000	001	030
Dunelmensis	001	001	005	004	062	011	020	000	000	021
Westmorlandia	000	000	004	006	026	008	015	000	002	019
Cumbria	001	001	008	015	058	020	033	000	003	008
Northumbria	000	000	011	012	040	021	016	000	001	008
Monumetha	000	000	006	007	142	015	014	001	000	008
Glamorgan	000	001	007	012	151	016	006	000	000	005
Radnor	000	000	004	005	043	013	005	000	003	000
Brecknok	000	000	003	004	070	027	013	000	000	002
Cardigan	000	000	004	000	077	026	009	000	004½	000
Caermarthm	000	000	006	004	081	020	016	000	004½	002
Penbrok	000	001	006	005	142	006	007	000	002	003
Montgomer	000	000	006	003	042	028	006	000	000	000
Merionidh	000	000	003	002	034	026	007	000	000	000
Denbigh	000	000	003	003	053	024	006	000	000	006
Flint	000	001	003	004	024	004	002	000	000	002
Anglesey	000	000	003	000	083	008	002	000	000	000
Caernaruan	000	001	005	003	073	017	006	000	000	000
Shires 52	25 Cities	25 Bysho.	641 Marto.	182 Castl.	9725 Parch.	554 Rius	256 Brid.	11 Chase	66 Forr.	781 Park.

29

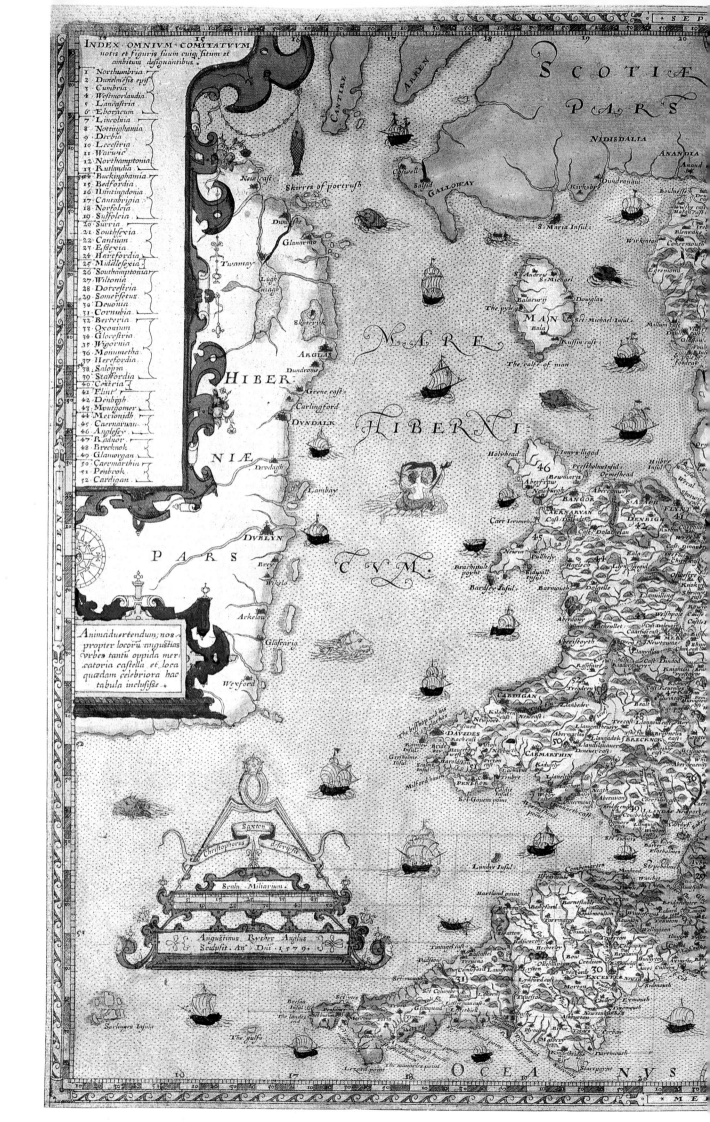

SCOTIÆ PARS

INDEX OMNIVM COMITATVM
notis et figuris suum cuiq; situm et
ambitum designantibus.

1. Northumbria
2. Dunelmensis epis̄
3. Cumbria
4. Westmorlandia
5. Lancastria
6. Eboracum
7. Lincolnia
8. Nottinghamia
9. Derbia
10. Lecestria
11. Warwic
12. Northamptonia
13. Rutlandia
14. Buckinghamia
15. Bedfordia
16. Huntingdonia
17. Cantabrigia
18. Norfolcia
19. Suffolcia
20. Surria
21. Southsexia
22. Cantium
23. Essexia
24. Hartfordia
25. Middlesexia
26. Southamptoniar
27. Wiltonia
28. Dorcestria
29. Somersetus
30. Deuonia
31. Cornubia
32. Berreria
33. Oxonium
34. Glocestria
35. Wigornia
36. Monumetha
37. Herefordia
38. Salopia
39. Staffordia
40. Cestria
41. Flint
42. Denbigh
43. Montgomer
44. Merionidh
45. Caernaruan
46. Anglesey
47. Radnor
48. Brecknok
49. Glamorgan
50. Caermarthin
51. Penbrok
52. Cardigan

HIBER-

NIÆ

PARS

Animaduertendum; nos
propter locorū angustias
vrbes tantū oppida mer-
catoria castella et loca
quædam celebriora hac
tabula inclusisse

Saxton
Christophorus descripsit

Scala Miliarium

Augustinus Ryther Anglus
Sculpsit Añ Dñi 1579

NIDISDALIA

ANANDIA
Anand

GALLOWAY

MAN

MARE

HIBERNI

CVM

CARDIGAN

CARMARTHIN

PENBROK

OCEANVS

30

ANGLIA
hominū numero, rerumq̃ ferè
omniū copijs abundans, sub miz
tissimo Elizabethæ, serenissimæ
et doctissimæ Reginæ, imperio,
placidissima pace annos iam
viginti florentissima.

Anͦ Dñi
1579

OCEANVS

GERMA-
NICVS.

BRITANNICVS.

GALLIÆ PARS

Cales
Escalus
Newhauen
Bolonge

31

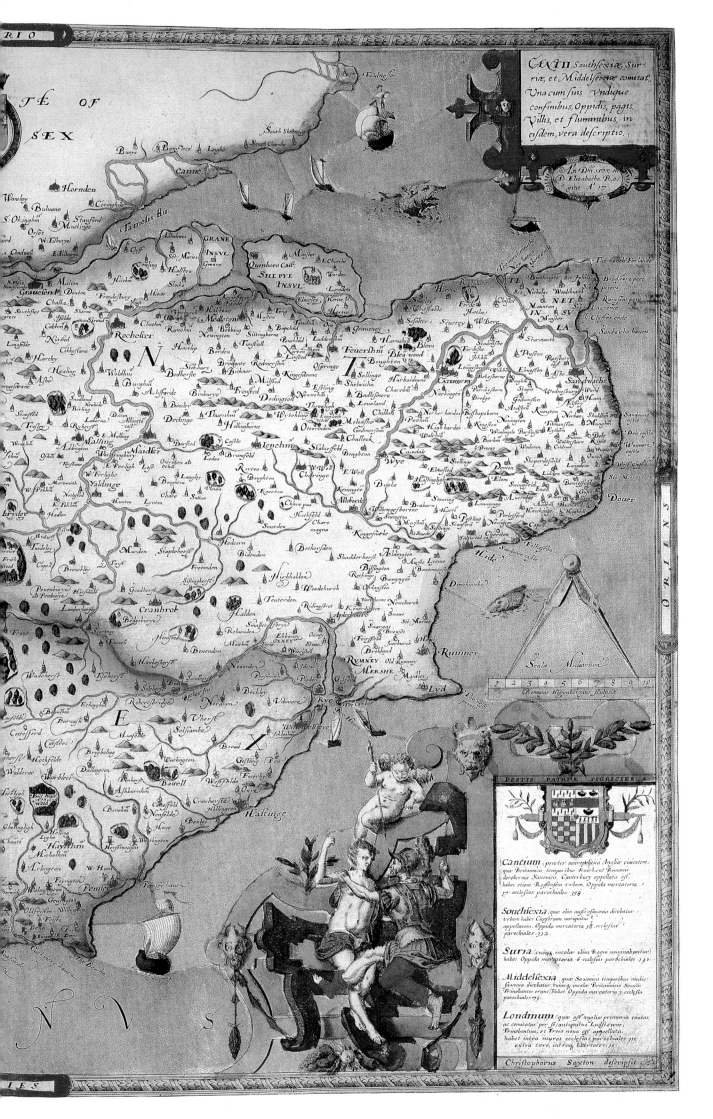

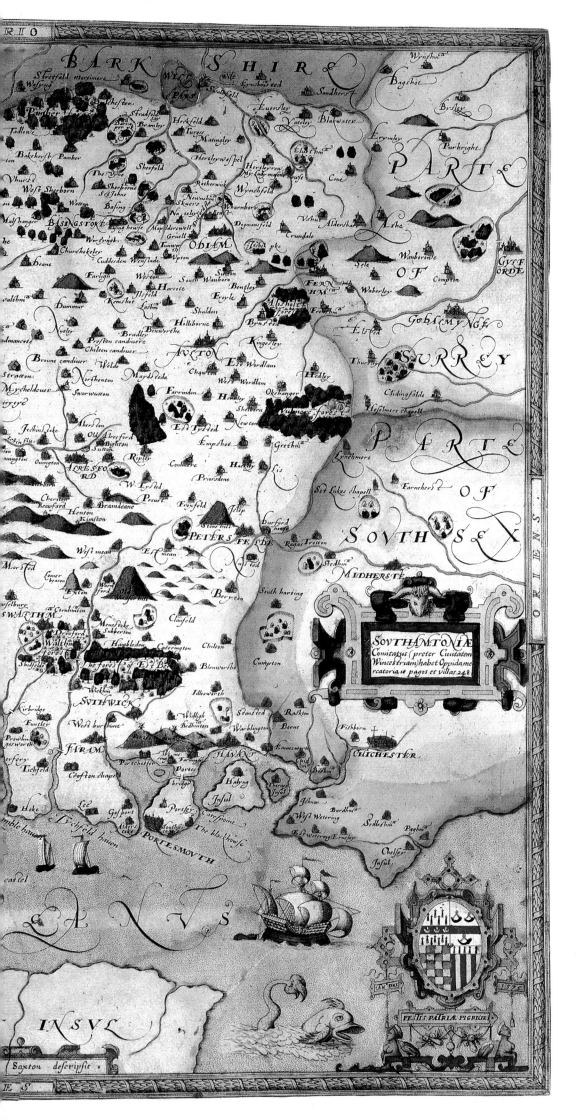

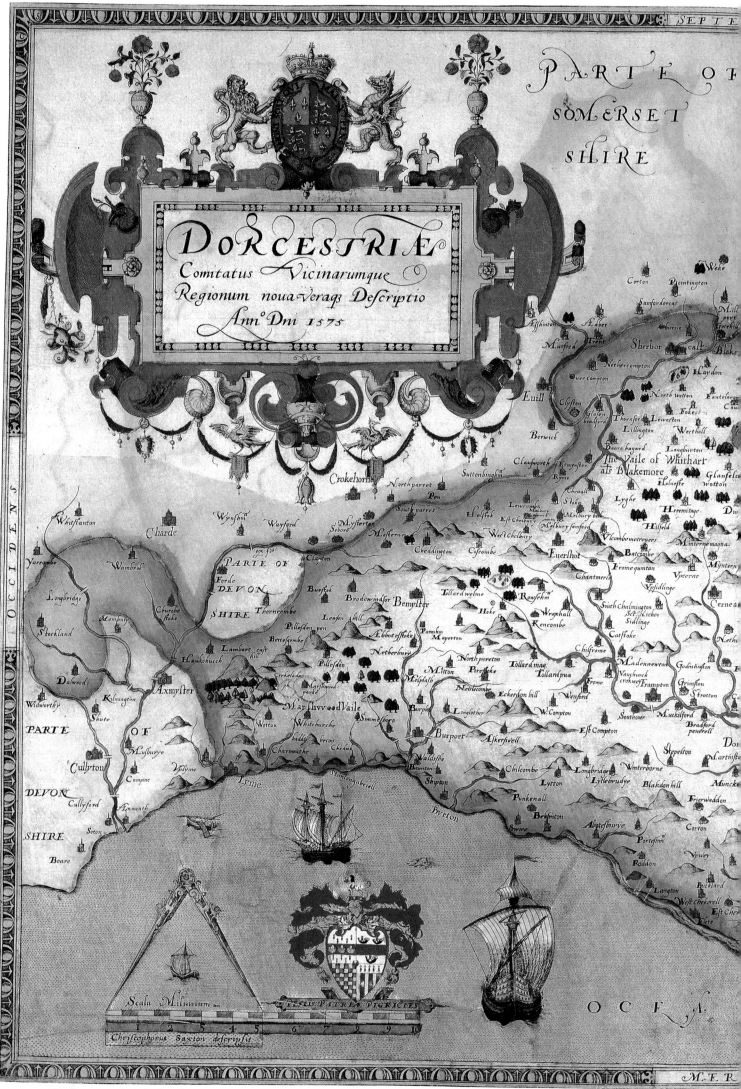

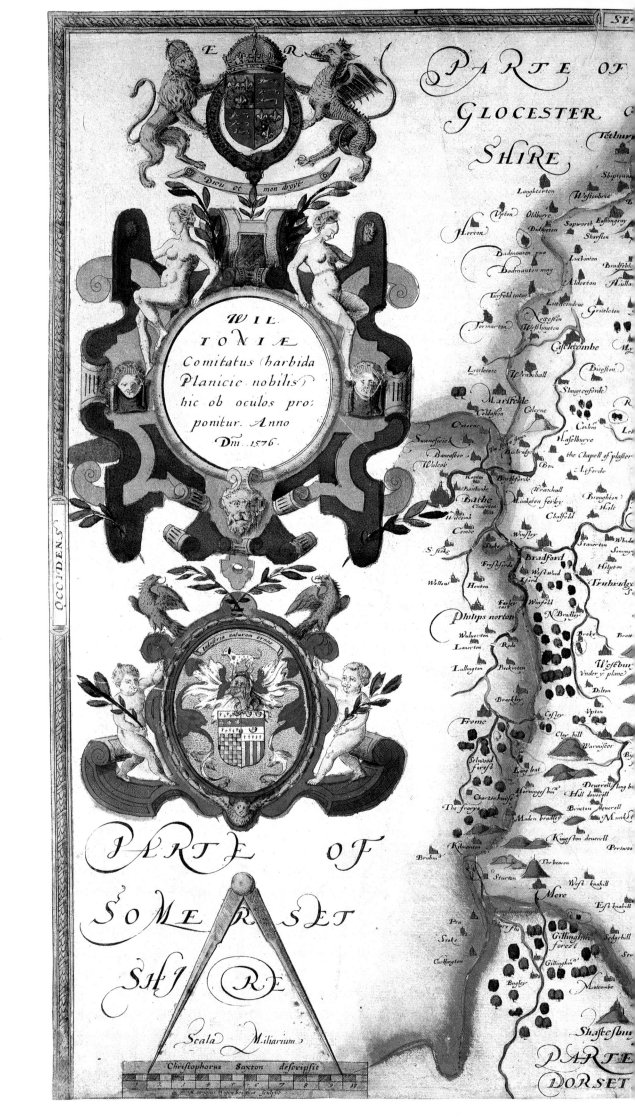

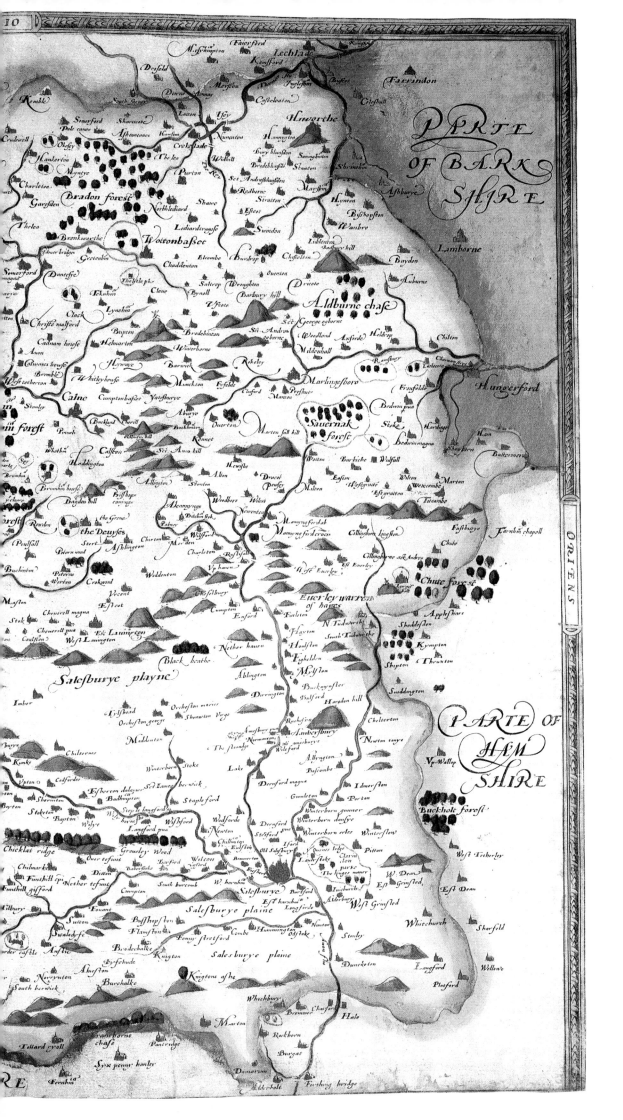

PARTE OF BARK SHIRE

PARTE OF HAMSHIRE

ORIENS

OCCIDENS

Flatholms Insul

Wes ton

Stepeholms Insul

Bray

Barre

Stert poin

SOMERSETENSEM
Comitat. (agri fertilitate
Celebrem) hec ob oculos
ponit Tabula

DIEV ET MON DROY

Anno. 1575. et D.
ELIZABETHE Regine A° 17

Batesfall point

Porlock bay

WATCHET

Stokland marshe

Kuline

Lysfoke

Ourg Culbons Porlok Selworthye
MYNHEAD

Stokland Otterhampton
Kilton Stremton Stokegurser Comidge
E. Quantoke heade Dansford Alsaxton Fiddington

Luccum Weston Curtenay
Stoke pero Timbercombe DVNSTER

Sct. Decombes W. Quantoke heade
Old Cleue Houlsford Nether

Exmore Corhinton Wellyton Doddington Cannington

Ex flu Exforde Cutcombe Luxboro Wichicombe
Nettlecombe Bicknaler Ourstowey Coripole Wemdon

Leighlande Combe Samsford

Trehoro Monkesiluer Stoke gomer Crokom Charlynche
Baske flu Wethipole Winesford Exton Wechihill Elworthye Spaxton

Brunton regis Laurandediarn Bagboro Brumfelde Enmore Gotehurst

Dunsbrok flu Haukeridge Wpton Hewishe Clatworthye Combfleorye North Pether

DVLVERTON Haddon beacon Tollande Combfleorye Kyngston

EstAustie Bittescombe BVSSHOPS LEDIARD Thurlxton

Tuchen Comba Skilgate Chipstable Erfs heade Haulse Hestercombe Monketon

Molland WestAustie Butsfsford Murbathe Ralington WIVELESCO Hethefeld Norton Staple Cheddon Cre

Langridge Dysford Petton MILVERTON Oke grove

Exbridge Clayhanger Stawley Hilfarrens Helebridge Rysson
Highley BAVNTON Baddel ton Nynheade Hilhushep TAVNTON T

Ashbrittell Langford Bradford Wilton Mary Stoke

Holcombe rogue Rittesford Ruyton WELLINGTON Trull Orchard Whatche

PARTE OF West bucland Pounsford Thurlebare Hatche

Margrest hurne Samsford Angefley Pitmyster Corse Bicknell

Canons hyghe Buckland Blak don hills Bapton Staple

DEVONSHIRE Church staxton Oteford Roche cast hill Ne

Clay haydon

Christophorus Saxton descripsit
Scala miliarium
1 2 3 4 5 6 7 8 9 10
LEONARDUS TERWOORT INTVERPIAN: SCVLPSIT

Yarcombe Whis

Wambroc

Longbridge

Stockland Meml

42

PARTE OF SOMERSET SHIRE

Perlock   Schwerthy

Colbone

MORE   Wachipole   Wynesford
Huckeridge   Duluerton   Bittescombe   Chipstable
Brushford   Radington
West Anstre   Holybut   Stawley   Kitresford   Raxton   Wellington
Langridge   Stolgule   Peten   Samferdarrundell
Knawston   Olcford   Baunton   Clayhanger   Abbriuels   Margrostharms   Ampton
Warnostlay heco hill   Studley   Hamtesham   Holcombe regus   Otterforde
Cron   Hucke   Caue   Commons   Buckland
Roeashe   Loxbeure   Mere   Naleman   lygho   Burlescobe   Columbdaund   Churchstanton   Whitstanton   Creket
Rakenforde   Wassfelde   Sanfordpeucrel   TScolumbe   Clay hayden   Wackpedowne   Warcomb   Charde   Wurstin
Myll   Chistcombe   Ascford   Hemnyek   Yarcomb   Wambrok   PARTE
Temple   Teuerton   Haulberten   Welland   Broadfolde   Shildon   Upautre   Ronesaucry   Stokland   Hemburye   Churche   Forde OF DE: VON SHIRE   Thornecome
Cadley   Columbton   Woode Athelewe   Luppe   Hunton   Dalwood   Kithuynyton   Azmilter   PARTE OF Hawkechurche   DORSET
Bickley   Butterley   Mora   Rontgshore   Old dunkeswel   Hunyton   Offwell   Waiders   Shute   Drake   Froneswes   Lyme   SHIRE
Bradinch   Silueton   Brodehembury   Combere   Ceeley   Colyfarcll   Sayton   Colliton
Neroey   Beare   Pehembury   Ferham   Sa: Michaell   Combe   Merly   Culliton   Southley
Thorueton   Reve Stoke   Chst Cayden   Owolsecomk   Noely   Farwey   Comuns   Iplym
Upton pyne   Columb sche   Lauranes chse   Chst Ro: But chercell   Gresson   Spton
Hayne   Chst Bro: Tillam   Crawton   Holcomb   the cob
Pines   Wymple   Cadde   Autre Sei maries
Newtonsyres   Polly mora   Brde Chyst   Strote wood heod   Sydbury   Branscomb   Boare
Oldridge   Pyn hawes   Hunyton Chsse   Harfarde   Salcombe   Bransworth point
Whitston   Rockbere   Newton
Hetherton   EXCESTER   Swimb   Aleshere   Zenautbre   Sidmouch   OCEANVS
Sei Maryedborne   Sei Thomas   Iford   Bushym   Collaton
Holcombe   Ede   Exmpser   Ferwyaden   Woodbury
Dunsforde   Preamouts   Chst Sei Maries   Bicton   ORIENS
Dunsfedh   Shellingford   Chst Sei George   Anterton point
Dodescombelyob   Newforde   Powderham costel   Netwellcourt   Tibwel   Salcome
Bridforde   Aschton   Oxton   Wochcombehap   Wochicomus
Kirstewe   Trustin   Manbende   Slort poite
Chudliy   F: Sheombe   A: memohe   Chckstow
Hennock   Edeforde   Bulysha
Beutrosse   Kingstxt   Vghrock
Tuaggrace   Preston   Tenymouth
Knowssayms   Buffhopstaynten
Hewr   Combe in tone   Worthe stone
Uber
Bockuste   Newton bushel   Steke
Atgnell   A botsheswell   Hacthm
W: Weswell   Cffynsswell
Denbury   Kingesheswell
Bryan   Sspplepton   Apsham
Compton   Stert point
Merledon   Tirmouse
Bery east   Paynton   Tor bay
Brewe piscuye   Stokegabriell   Woston
Stokegabriell   Chesston   the bery poynt
Cornesforde   Aysham   Breysham   Greneweye
Dryslton   Besumston
Noshill   Sei Stone Cur   Kingesswere
Sct Stone Cur   Dartmoth   Trounte chapoll
Slupton   Monstone   BRITANNI
Scala Planeryca   Dartmouth hauen   The Combe point
Stert point   CVS

DEVONIÆ COMITAT, RERVMQVÆ omnium in eodem memorabilium re: ceus, vera pticulariq, descriptio. Anno Dñ 1575

Scala Miliarium
Christophorus Saxton descripsit
Remigius hogenbergius sculp

ES

43

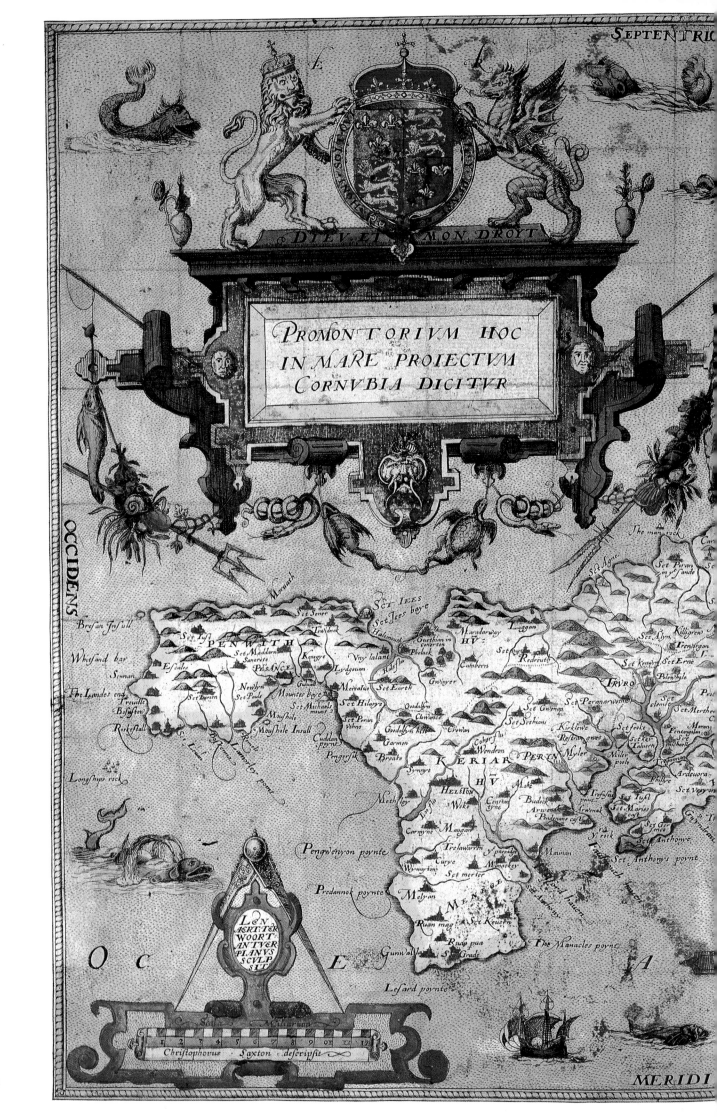

DIEV ET MON DROYT

PROMONTORIVM HOC
IN MARE PROIECTVM
CORNVBIA DICITVR

The man rock

St Agnes

St Piran
in y sande

Morvach

St Sever

St Iees

St Iees baye

Tewidnak

Halmouth

Guethian in
conerton

Phelock

Maçadarway

Luggan

HV

Bresan Insull

St Just

PENWITH

St Maddarn
Sanereete

PESANCE

Kengye

Vny lalant

Hale flu

Redreuth

St fort

St Lyn

Killigrew

Trengwgan

Whisand bar

Senan

Escales

Lydgewan

Gwinyer

Canbert

St Kenefy

St Erne

Palewhyle

The Landes end
Treuille
Bofuftow
Rorkeftall

St burien

Newlyn
St Paule

Gulnall
Mounts baye

Meretalue

St Enrth

Goodalf
Clovance

St Gwenop

St Peranarwothel

TRVRO

St clement

St Merther

Moushole
Moushole Insull

St Michaell
mount
St Pran
Vehno

St Hilarye

Goodalf-n hell

Crowan

St Sethian
Kirklewe
Reftonogwes

St feoke

St Ker
Taluern

Morā
Fentanggon
michaell

G. E.
Lunulter ponte
Pengerfu

Cuddan
poynt
Breake

Garmon

Synure

Cobar flu
Wendron

PERYN Myler

Miler
poole

Trefumus

Ardevora

Phillre

St Vezy on

Longfhps rock

Methley

KERIAR
HV

HELSTON
Weke

Constan
Eyne

Mabe
Budick
Arwenall
Pendenans cafle

St Tuff
Trefufus
point
Arwenal

St Maries
coff

St Ger
enes
St Anthonye

Pengwenyon poynte

Carnyno

Maugan
Trelawarrn Nparessage
Cupye

Wy-my-lans

St merter

Mmackey

Maunan

Yrock

Sct Anthony's poynt

Predannok poynte

Melyan

MENEGE

Ruan mag

St Keuern

The Manacles poynt

Gunwalla

Ruanpua
S Grade

OCE
A

Lefard poynte

| 1 | 2 | 3 | 4 | 5 | 6 | 7 | 8 | 9 | 10 | 11 | 12 |

Christophorus Saxton defcripfit

46

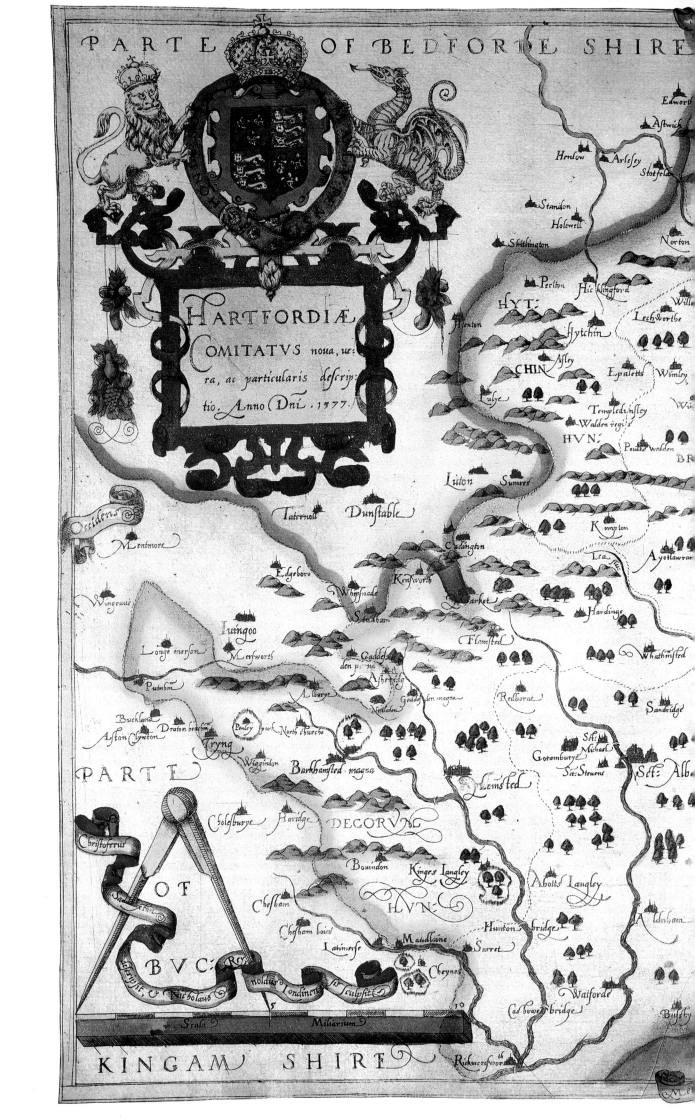

HARTFORDIÆ COMITATVS noua, ue: ra, ac particularis descrip: tio. Anno Dñi. 1577.

Occidens

Edworth
Astwich
Henlow Arlesey
Stotfeld
Standon
Holewell
Norton
Shitlington
Perlton Hichlingford
HYT: Willi
Lechworthe
Hexton Hytchin
CHIN Asley Epaletts Wimley
Eulye Templedinsley Wi
Walden regis
HVN: Poutes walden
BR
Liton Sunves
Kimpton
Menthnore Lea fici
Coddington Ayotlawrar
Edgeboro Kensworth
Whipsnade Market
Wingraue Stedham Hardinge
Taternoll Dunstable Flamsted
Lungoo Whathmsted
Longe merson Merfworth Caddesden pua Redborne
Putnhm Ashredge
Albury Gaddesden magna Sandridge
Buckland Neetleton
Aston Clynton Draton beaucm Penley park North churche Scc:
Tryng Michael
Wigginhn Barkhamsted magna Gotamburye Sct: Alb
Lemsted Sct: Steuens
Cholesburye Havidge DECORVN
Bonindon Kinges Langley
Christoferus Abotts Langley
Bonindon Aldnham
Chesham HVN:
OF
Chesham boies Hunton bridge
Latimerse Maudlaine Sarret Watforde
BVC: Rex Cheynes
Cashoure bridge Bughy
noldus Londinete fit sculpfit
Scala Miliarium
Rickmersvor th

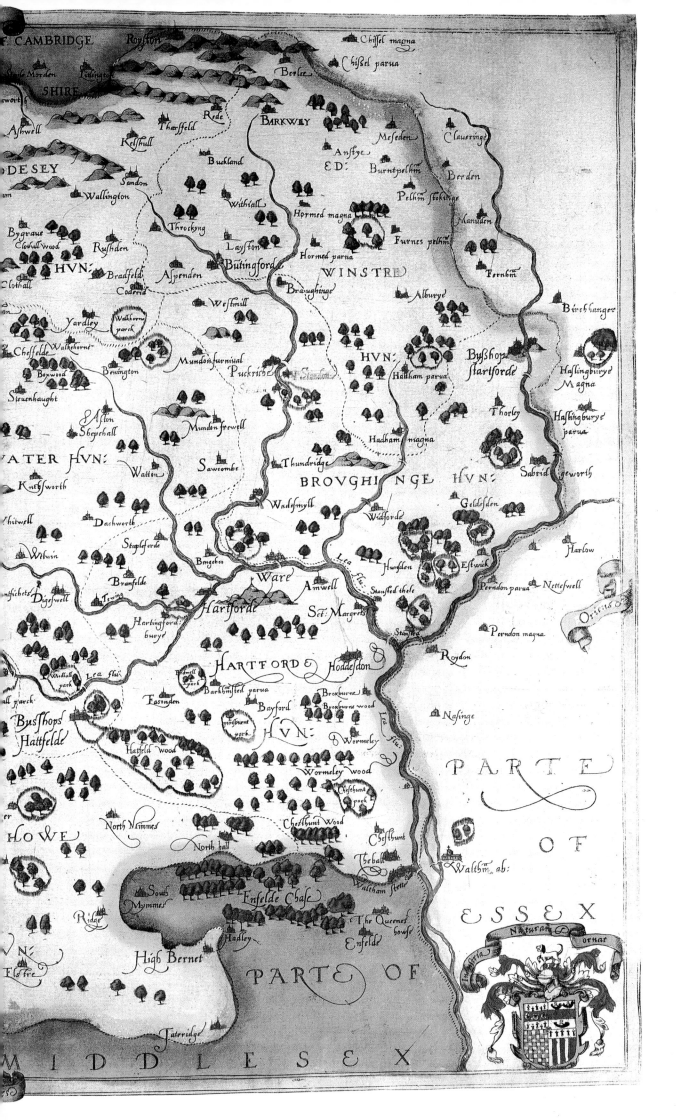

F. CAMBRIDGE   Royston   Chissel magna
Chissel parua
Morden   Lillington
SHIRE   Berlee
Ashwell   Rede   BARKWEY   Meseden   Claueringe
DESEY   Tharsfeld   Anstye   ED: Berden
Kelshull   Buckland   Burntpelhm   Pelhm stokinge   Manuden
Sandon   Withiall   Hormed magna   Furnes pelhm
Wallington   Throckyng   Furnes pelhm   Ternhm
Bygraue   Layston   Hormed parua   WINSTRE
Clothall wood   Rushden   Butingford   Alburye   Birchhanger
HVN: Bradfeld   Aspenden   Braughinge   Byshops
Clothall   Coderid   Westmill   HVN: stantforde   Haslingburye
Yardley   Walkehorne   Mundon furniuall   Hadham parua   Magna
Chessfelde   Walkehorne   park   Puckrich   Thorley   Haslingburye
Boxwood   Benington   Staxton   parua
Steuenhaught   Aston   Mundon frewell   Hadham magna   Sabridgeworth
ATER HVN: Shepshall   Saweombe   Thundridge   BROVGHINGE HVN:
Knebsworth   Watton   Wadesmyll   Widsforde   Geldesden
Whitwell   Dachworth   Husselen   Estwick   Harlow
Wilwin   Stapleforde   Bengehoo   Lea flu: Perndon parua   Netteswell
Bransfelde   Ware   Stansted thele   Oriens
Digeswell   Tering   Amwell   Perndon magna
Lea flu: Hartforde   Sct: Margrets   Stansted
Wickhall   Hartingford: Roydon
park   burye   HARTFORDE   Hoddesdon
Busshops   Easonden   Barkhmsted parua   Broxburne   Nasinge
Hattfelde   Bayford   Broxburne wood
    pymsburne   HVN: Wormeley   PARTE
    park   Wormeley wood
Hatfeld wood   Chesthunt   OF
North Nimmes   park
North hall   Chesthunt Wood   Chesshunt   ESSEX
HOWE   The ball
South   Enfelde Chase   Walthm ab:
Mymmes   Waltham stotte
Ridge   The Queenes
VN: Hadley   howse
Elstre   High Bernet   Enfelde
    PARTE OF

    Fatridge

MIDDLESEX

49

Oxonij buckinghamiæ et berceriæ
Comitatuum, vna cum suis vndiq̃
Confinibus, oppidis, pagis, villis,
et fluminibus in eisdem vera
descriptio. Anᵒ Dm̃ 1574.

Oxonium preter Academiam et villam
Oxoniensem habet oppida mercatoria 9
Ecclesias parochiales 208
Buckinghamia continet in se oppida
mercatoria n ecclesias parochiales 185
Berceria continet in se oppida
Mercatoria n ecclesiæ parochiales 139

PARTE OF BEDFORDE
SHIRE

PARTE OF
HARTFORDE
SHIRE

Set Albons

ORIENS

PARTE
OF
MIDDLE
SEX

THE VALE OF ALES BVRYE

PARTE
OF
SVRREY

Scala Miliarium

1 2 3 4 5 6 7 8 9 10

Christophorus Saxton descripsit

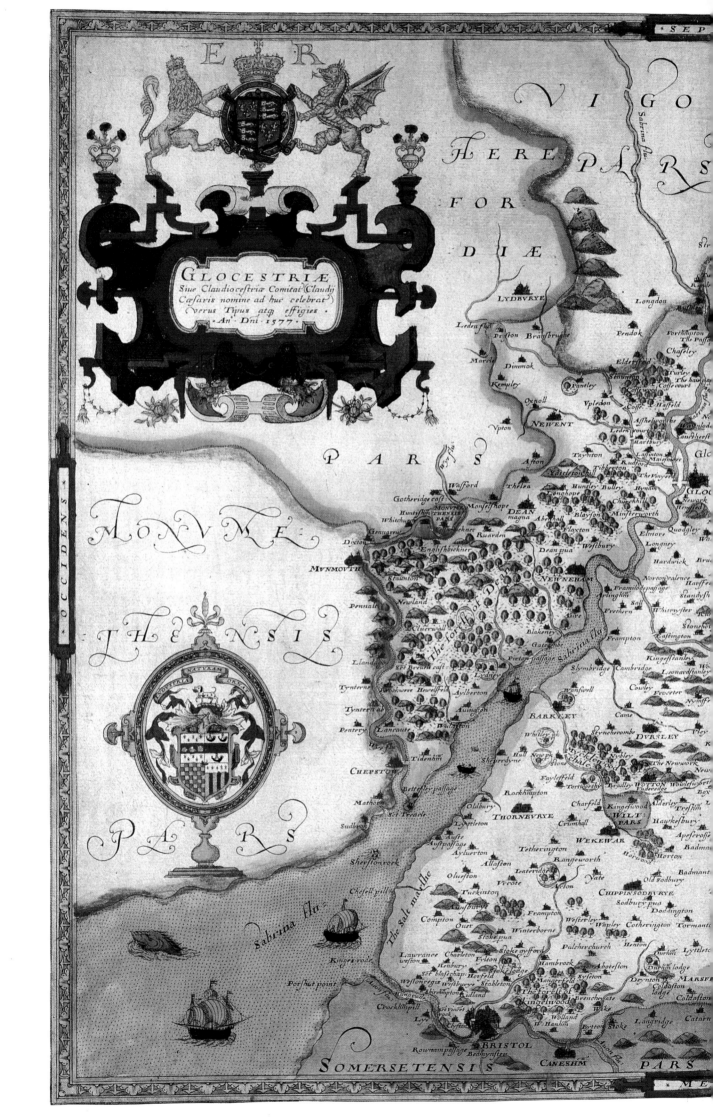

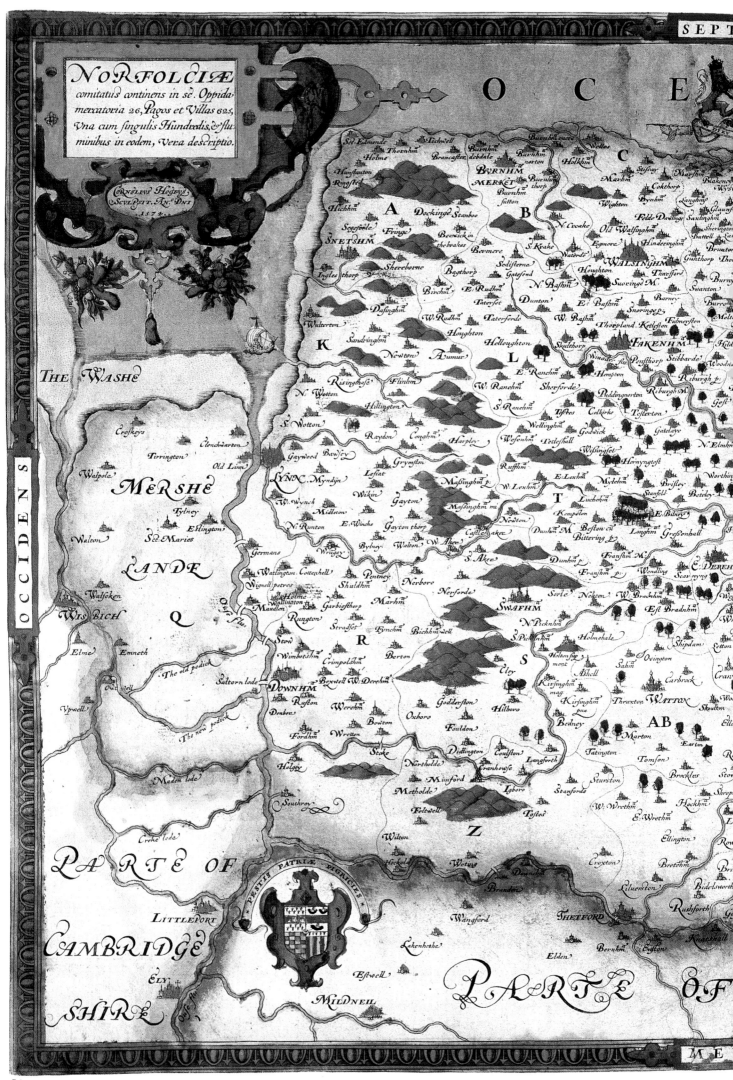

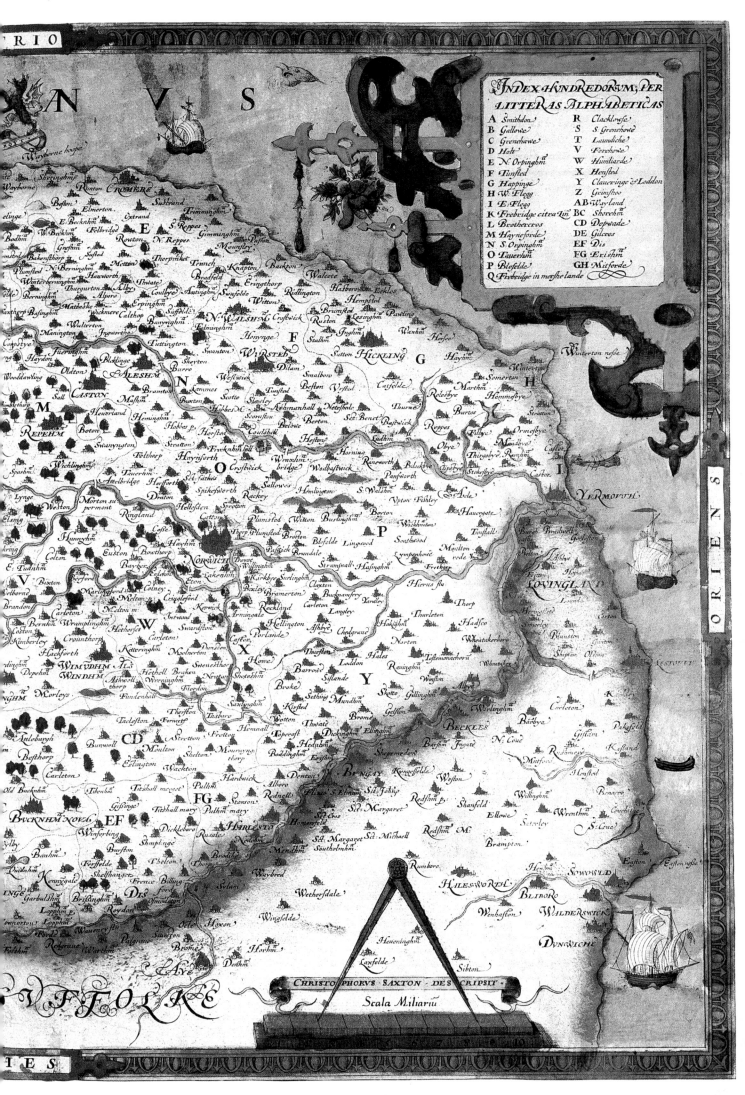

WARWIC=
LECESTRIAEQ3
Cōmitat: Cuitat:
Oppidoru Villaru.
fluminu Ceterarumq3
rerum omnium in
eisdem memorabi:
lium. noua Veraq3
descriptio.

OCCIDENS

PART OF
SALOP
Hales ouen
SHIRE

PARTE OF

WORCESTER

Industria naturam ornat

SHIRE

PESTIS PATRIÆ PIGRICIES

PARTE OF
GLOCESTER
SHIRE

LYCHFELD Whittenton
Fishrick Elford
Comberfor
TAMEWORTH

Berne flu: Weford
Shenston Hynte
Amyn
Villinco
Drae
Aldridge baffet
Rushall Myddleton Hole
Sutton cofeld Kinburye
Barr Newhall Moxhall
Hamsted Wyfmawe
Sandwall Peryhall Curdworth Leo Nether Who
Handsworth Yarneton hall Tame Castlebramiche Waterorton Shifftoke
Aston Makefoke caffe
Ouldbury Kingfhurst COLESHIL
Smethik Dudson hall Yardley Makefoke
BROMYCHM Sheldon Packinton p
Edgebafton Packinton ma
Horborn Helmedon Bykenhull Packinton
Rea flus Mowfley SOLYHUL Hampton
Kyngesnorton Heluoon Barſon Barkeswell
Cofton Longdonhall Riffon
Knoll The temple of Thе
Arrow flu Wrtho chapell balshall Quen erk
Aleburche Belcs W I Packwood Hunnley
Tardbik Burfley Skiltes Nuthurst Baddefler Nether
Heuedgrange Ipfley Tanworth Lapworth Rawhall
Feckenhm Apfley Rowinton Haſeley
forest Morton Outnall Prefton Hatton Groue Budb
Feckenhm Oulbarton Cloverdon Pylter Hampton
Studler Sparnoll HENLEY Waluar dington Norton
Coughton Aune lodge Edfon Sherborn
Benchm court Afen S Woston Bearley Smitorfeld Bearfou
Inkbarrow Arrow Kynnerton Afton Wafburton
Ragler Edfron Buffhophinton
Morton Wetheley AYLCEST ER Hofeler Clayton
Roufelmche Benwinton Morehall Grafton Buffhopton Auſton
Abbot merton Bixford Bruton Draton STRETFORD Walton
Sanford Brome Londonton Loxley Golder
Haruyngton Clelie Wefard Wilote Prefton R
Auon flu Doffy Afton Aldermarfton
Carleton Uffenton N Lytleton Polwoth Marton fickwo Ouer eatenton Butlerfmar
Brode marfton Quernton Eatenton Nether Piller
EVESHOLME Newbo Hawforde Ouer
Myckleton Stene Blackwell Ecchddinton Whatcote
Thornton Hunnyton Iddlecote Lyfter
CAMDEN SHIPSTON Compton
Ebberton Tidmyng Winderton
Parford Stratton Burmyngt
Worces: Hangryngton Cherinton
Blockley ter Tredyngton Wudford
shire
Monton hynmarfhe Barton on Whichford
the heath
The foure shire ftones Longe compton
Eueflode
Worcefter Oxforde Glocefter MER
shire

60

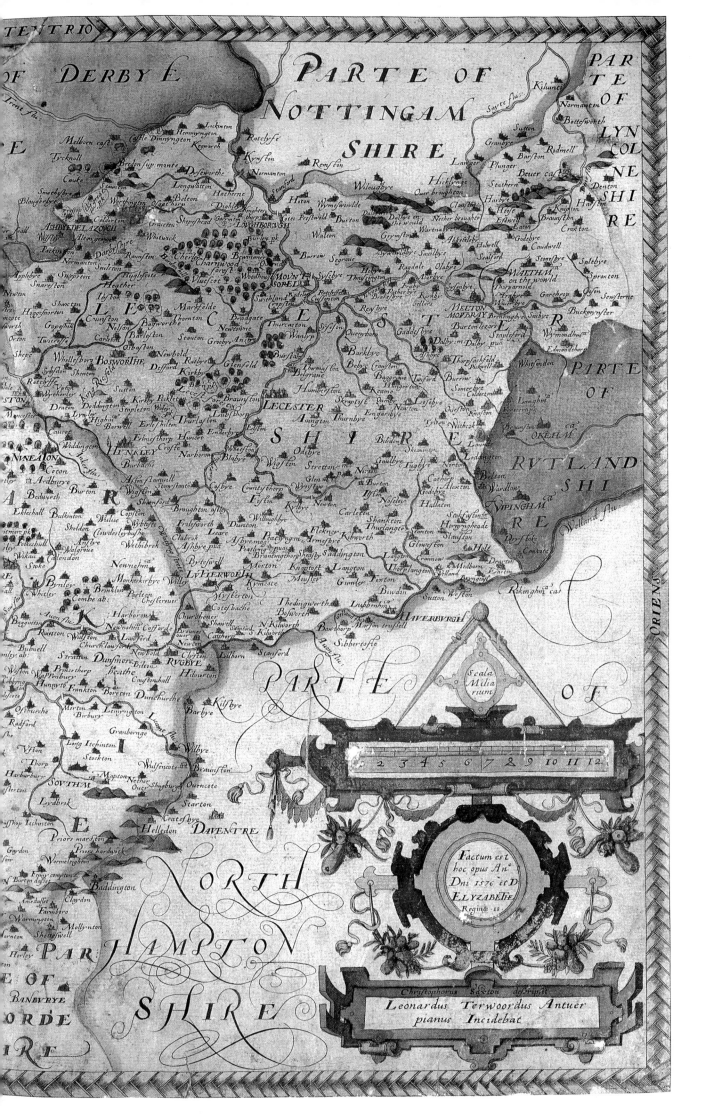

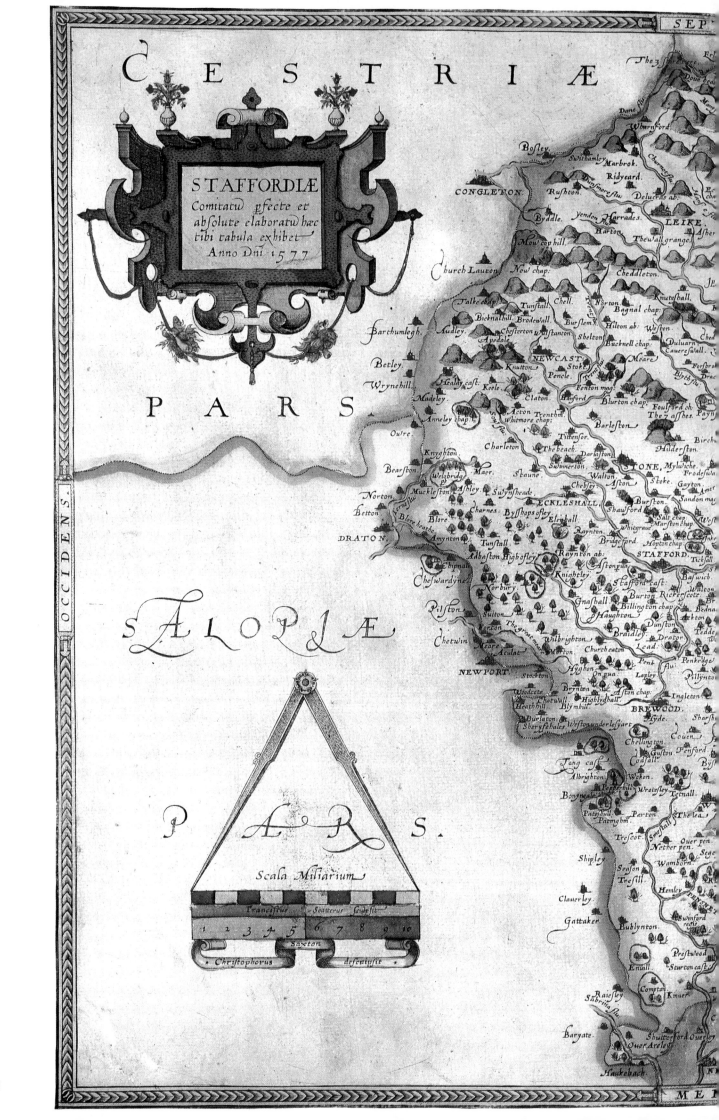

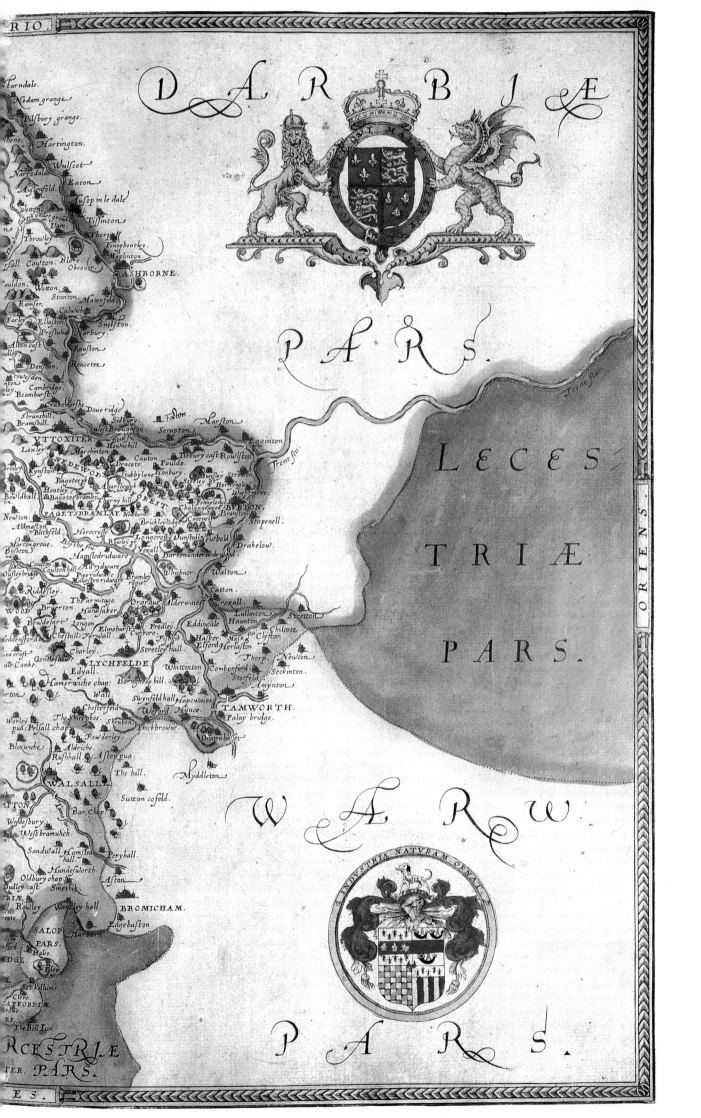

Spaldinge · PARS · NORFOL: ·

COLN: · Set Maries tide

Coubet · Guyth

Crowland · Dowesdale · Clowes croffe · Newt · Walton

Merkerdeping · Leuerington · Walfoken

Thekenhold · Sandroue · Emneth

Set Maries · Wilbiche · the old podick

Peakirk · Thorney · Gherne · Elme · PARS

Thornewaie · Outwell

Eye · Oxney · Hebbes howfe Coldam · Vpwell · Salters Lode · Downeham

Peterbrugh · Waterfes leame · The new podick · Denber

Whittlesey · Merfh · Welney · maden lode · Helger

The new leame · Southrey

Whittlesey dike · Wimlington

THE ISLE · OF · ELY

HVNTING · Yaxley · Whittlesey mere · Ramfye mere · Dummyngton · Prefthoufe

Ramsey · Littleport · Shreye · SVFFOL:

Chatres · Downeham · Copthall · Mildnall

DONIA · Mepole · Coueney · Newberye · Prickwillow

Bury · Wichford · Ely · Stantney · Iflam · Wormleighton

Wentworth · Bedesham · Brame · Norney

Set Andras chap: · Thetford · Hardware chap: · Sohm · Fordham

Sutton · Sohm · Badlingham

Hadnam · Wilberton · Strethm · Wiken · Snalewell · Chipnhm

Wilberton · Kranet

Audre · Long Fen · Stuchsworthy · Vpware · Lanworthe · PARS · Kenford

Denny · Burwells · Exnyng · Moulcon

Cotenhm · Reche · NEWMARKET · Afhley

Ramton · Water beache · Swafhm priory · Ditton · fauxton

Willingham · Land beache · Swafhm bulbek · Newmket · Chefterf

Swafye · Horwynfeye · Botffm · Heathe · Stachworthy

Longftanton · Myleon · Stow qui · Weftley · Brinklow · Catlidge

B · H · Yftons Jmpinton · Wilborhm pua · Conhge

Lowlworth · Heggnton · Girton · Ditton · Wilborhm mag: · Bradley

Knapwell · Drydraton · Coton · Teuerfhm · Dullynghm · Burrough

Elfworthe · Childerfley · Madingley · Cherry hynton · Fulbornes · Willingham · Carleton

Boxworth · Hardwick · Berten · Grancefet · CAMBRIDGE · Gogmagog hills · Wefton cowell · Witherfell

Croxton · Tofte · Comberton · Trumpington · W'wratting · Thirlow

Caxton · Cawcote · Haffingfeld · Shelford m: · Stapleford · Buffm · Wickhm

Buene · Enerfden pua · Hawkfton · Shelford pua · Badburhm · Abbingeon · Horfheathe

Kingfton · Euersfden magna · Harlefton · Sawften · Hilderfhm · Hauerill

Gamlingay · Wymple · Harvleton · I · Witlesford · Pampworthe

Trowefley · Harley · Barrington · Widford · Foxton · Newcon bridge · Hinkefton · A · Lynton · Shdre campes

Gamlingay · Arrynton · Orwell · Malcon · Sheporheathe · Duxworthe · Hadftok · Caftle campes

Cokinharley · Harley magna · Crawden · Fulmere · Ickleton · Chesterfordes

Porton · Wormleghton · Clopton · Wendye · Whaddon · Meldred · Triplow · Strethall · Elmeden

Sutton · Sandye · Gransden · Shenye · Molborne · Crefhall · Chiftall magi

Eworthe · Tadlow · Gilden morden · Abbington · Basingborn · Royfton · Berlee · Chiftall pua

Biglefwade · Stratton · Duncon · Stepfmonden · Lytlyngton · Reade · ESSEXIÆ

Wilfumftead · Haunes · Southbill · Longford Edworth · Herneworthe · Ashwell · Kelfhull · Tharffeld · Barkwey

Shefforde · Chikfand · Aftwick · Cawcot · Newnhm

Chifton · Henlow · Arlofey · Radwell

Mauldon · Clophill · Cambleton · Stotfold · Norton · Baldock

Authill · Wrafte · Graueneroftes · Standon · Hicklingford

IA · Weftyng · Berton · Periton · Hitchin

Harlyncton · Sharpenho

HARTFORDIÆ

Tuddington · Streteley · Hexton · Sundon

Chalgraue · Chaulton · Bifcot · Lygraue · PARS

Tilefworthe · Houghton · Summes · Waldenregis · Luton

Eaton · Dunftable · Chaddington

CHRISTOPHORVS · SAXTON · DESCRIPSIT

Whipfnade · Keyfworth · Strethm · Merket · Scala Miliarium

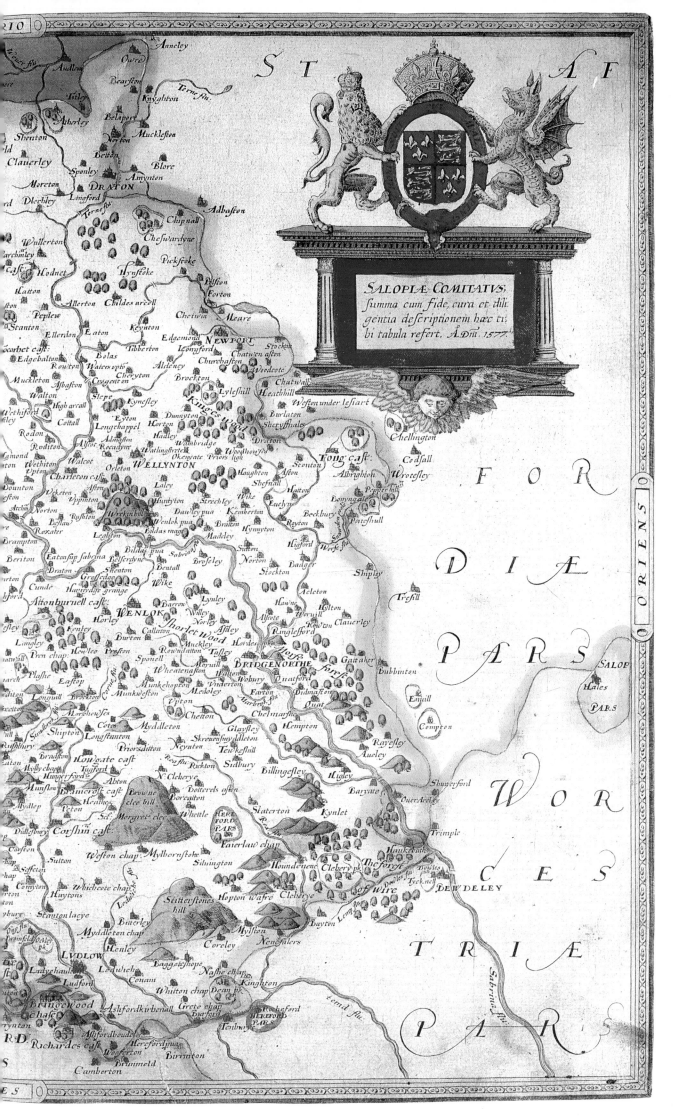

SALOP**SÆ**

Corve Riu.
Oney flu.
LVDLOW
Auldon
Marlon
Ladyhaulton
Kynnen
Moektre
Ludford
Winen
forest
Treppleton
Walfork
Lanewardyne
Don fare
Brugewood
chafe
Terne flu.
Buckhall
Buckton
Burrinton
Hereford
Bragfeld flu.
the grange
Atforton
Afton
Afhford chutchy
Teme flu:
Elton
Richardes coppice
Woofterton
Bri
Scandyfhe
Brantenbryan caft
Brimmeld
Pedderton Leiton
Wigmore caft
Camberton
Lug flu:
Lytton
Derefolde
Lentall ftarke
Orleton
Afhton
HEREFORD
forest
Ouer lee
Lentallerles
Morton
Mydle
PARS
Lyngan
Nether lee
Yeaton
Birchall
Birriton
Lymbrok
Lug flu.
Croft caft
Yarpull
Eye
Difcoyde
Kynftan
Pyton
Shirley
Aylmyftre
Luckton
Lufton
Stockton
Stepleton caft:
Afton
Kinr
PRESTAYNE
Conye
Eyton chap
Kingefland
Eaton B
oney. flu:
LEMSTER
Nafhe
Shobden
Chorleftre
Stanbache
Lawton
Stogbache
Bryverley
Knyll
Titley
Areefland
Euington
Staunton
Munkland
Warton
Arre flu:
Noke
Burton
RAD
PEMBRIDGE
Lenals caft
Stretford
Ouerhill
Newton
Marfton
Delfin pua
NOR
KYNETON
Wefton
Delfin magna
Bearley
Wynfley
Hargaft
Broxwood
Lattons
Sarneffeld
WEBLEY
Peonregis
Wonton
Wellu
Glaftre
Almelee
Sarneffeld
Cannonpeon
Huntyngton
coffen
Bullynghm
Kynnerfley
Wormefley
Brilley
Erdefley
Norton
Carfop
Brinfop
Mychaelchurch
Willerfley
Nafor
Monfillacye
Burfel
Whitney
Winforton
Leiton
Brobere
Byffhopton
Crednell
PARS
Stanton
Kenchefter
Manffeldgamedge
Stretton
Wye flu:
Monyngton
Bridgefales
Hunting
Bettus chap:
Myddleloyte
Bredwerdyne
Mockas
Byford
Suggas
caft
By
Clyfford caft
The bache
Blakemere
Prefton
Eaton Clayh.
Hardwick
Dorfton caft
Madley
Aly
The Gulden vale
Fowemynd
Snowdell caft
Tibberton
chap:
Peterchurch
Hongaft
Fowchurch
Kingfton
Pofton
Arkefton
Wye flu. HAY
Turnafton
Monnynton
Druxton
Kewfop
Fowemynd
Chaunfton
Dudle
Crafwall chap:
chappell
Morehampton
Sang
Newcourt
Wonmebridg
Set Margretes
Howton
Bacton
Llanyhangleefkle
Dowre
Kenderchur
Harlefewas
BREKNOK
Llamhangell
Elftones brid
The old court
Dewlas
Longtowne
Roulfton
Kyned
Cledok
Llangua
Llanfillo
Llaneuny
Munno flu:
Gryfemond caft
The oldtowne
Trewn
HEREFORDIÆ
MONV
PARS
Michaelchurch
The fothok
Cumyoye
crucorne
PA

PARS
Scala Miliarium

Chriftophorus Saxton defcripfit

| 1 | 2 | 3 | 4 | 5 | 6 |

Remigius hogenbergius.

LANCAS
TRIA PARS

CESTRIÆ

PARS

UNIVERSI
Derbiensis Comitatus
graphica descriptio
1577

STAF
FOR
DIÆ

PARS

Scala Miliarium

1 2 3 4 5 6 7 8

Christophorus Saxton descripsit

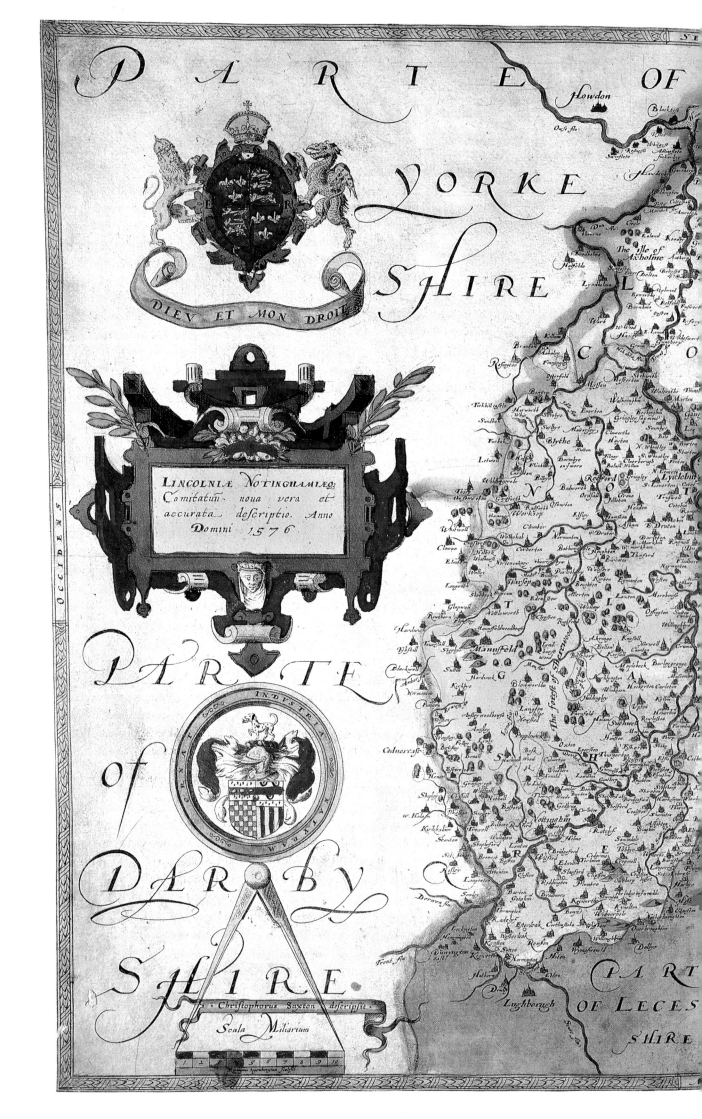

PARTE OF YORKE SHIRE

PART of

DIEV ET MON DROIT

LINCOLNIÆ NOTINGHAMIÆQ;
Comitatuū noua vera et
accurata descriptio. Anno
Domini 1576

INDVSTRIA ORNAT · INDVSTRIAM

PARTE of

DERBY SHIRE

Christophorus Saxton descripsit

Scala Miliarium

OCCIDENS

Place names on the map include: Howdon, Blacktoft, Swinsslete, The Isle of Axholme, Hatfeld, Lyndholme, Epworth, Finingley, Rossetou, Tickhillcastle, Sandbek, Blythe, Letwell, Worksop, Whitwell, Clowne, Bolsover, Hardwick, Glapwell, Blackwell, Hucknall, Mansfeld, Sutton, Kirkby, Codnor cast, Skegby, The Forrest of Sherewood, Nottingham, Radcliff, Kirkhalam, Stanton, Sandiacre, Kegworth, Normanton, Wymeswould, Loughborough, Dalby

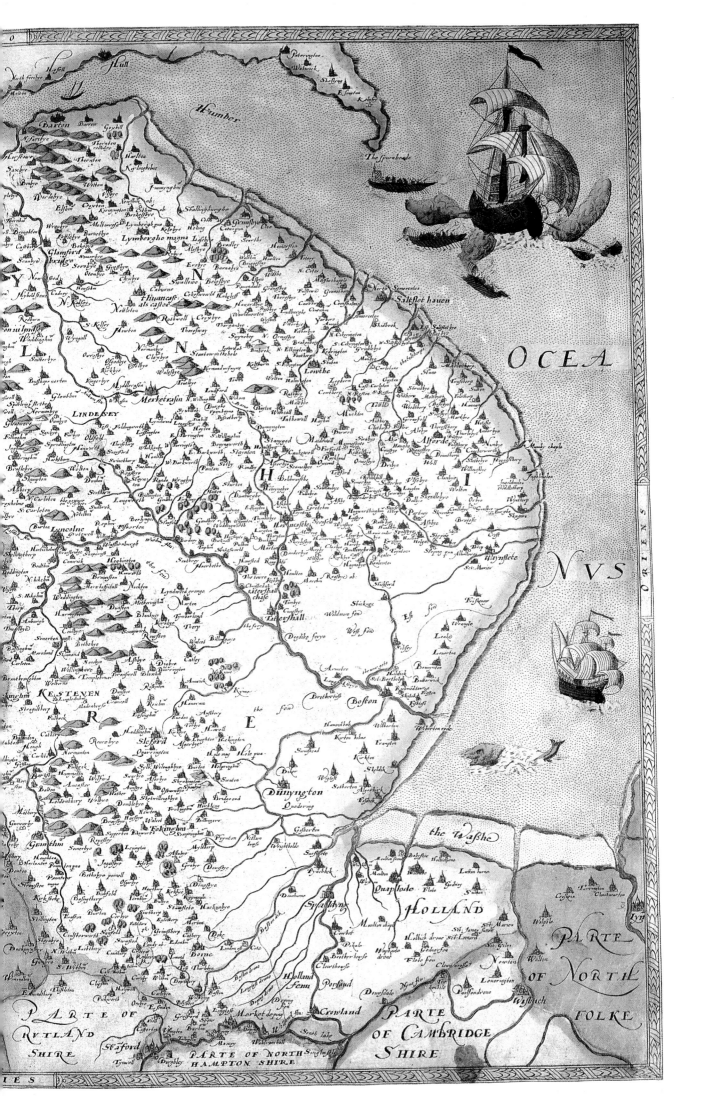

OCEA

NVS

ORIENS

The Spurnheade

Hull

Humber

Barton
Barrow
Goxhill

Newfar bye

Saltflet hauen

Y

L

N

LINDESEY

Lincolne

KESTEVEN

R

E

Tateshall

Boston

Donyngton

the Waſshe

HOLLAND

PARTE
OF NORTH
FOLKE

Wiſbiche

PARTE OF
RVTLAND
SHIRE

Staford

PARTE OF NORTH
HAMPTON SHIRE

Market depyng

Crowland

PARTE
OF CAMBRIDGE
SHIRE

LAN
CA

OCEA

NVS.

CESTRIÆ
Comitatus (Romanis
Legionibus et Colonijs
olim insignis) vera et
absoluta effigies.

OCCIDENS

Crosbye pua
Crosbye magna

Leuerpoole hauen
Bowtle
Bankehall
Kirkdale

LEVERPOOLE.

Wallasee
Poton
Sencombe
Byðston
Claghton
Tramnole
Morton
Vpton
Oxton
Saw conmasse
Newton
Ourchurch
Graiesley
Woodchurch
Prenton
Ouerbebbinton
Bebbinton
Westkirkbye Parkbye
Lancon
Stourton
Cauda Irbye Thingoo
Brinston Brumbro chap:
Branston
Oldfeld Thornton
Heslewall
Gayton Rabye
Leighton
Neston mag:
The newkeye
Nesse

Toch seath ph
Wartre

Farnworth
Appleton

Garston
Speakehall
Hut
Hale
Halewood

Stanley Ince
Estham
Hooton
Childeruthornton Poole
Willaston Suttoncourt
Whitbye
Stanney
Stoke
Croughton
Wyrwir
Pyckton

Cuerdley
Penketh

Merse flu
Runkhorn
Morton
Norton
Ouer Runkborn Haulton cast:
Weston
Rock sauage
Frodsbm
The cast
Netherton
Woodhowses
Helshy ton
Helshye
Elton
Thornton
Hapsford
Auonley chap:
Dunhm Manlay
sup mont

Bewsey
WARINGT
Sankye
The pele
Nether
Walton
Aston grange
High el
More
Kekewick
Dareshury
Hatton
Preston
Stockhm
Sutton
Grimsdic
Aston chap: Aston grange
Powsey als newbrug
Wuer flu: Dutto
Croton
Newton
Kyngeley Norley
Cuddinto
DELAMERE
FOREST
The chamber in the forest
Mouldesworth
The pyle
Budwo
Oulton

Balshark ab:
Holiwell
FLYNT.
Dee flu:
Burton
Puddington
Shotwick
Shotwick cast:
Mullynton
Blaconhall
The baites
Newton
Howl
CHESTER.
Christleton
Boughton

The lea
Capenhurst
Bateford
Charlton
Upton
Traford mag:

Bridgetraford
Plemestoo
Gildensutton
Ashton
Teruyn
Hocknel plat
Kelsale
Cotton
Duddon
Stapleford
Rowton
Warton
Staffeld
Utkinton
Darley
Clotton
Torperley
Burton

Barro
The pyle
Ashton
Kelsale
Flaxyeardes
Beston
Teerton
Bunbur

Ledsum

Yeuley cast:
FLYNT
Harden cast:
Broughton
Bretton hall
Eccleston
Dodleston
Eaton
Pooton
Kynnerton
Burton
Powsford

Huntington hall
Soghton hill
Churton heath chap:
Lea hall
Aldford

Hatton
Hakesley
Totnall
Golborne
Hanley

Beston cast:
Peckforton
Ridley
W

INDVSTRIA NATVRAM ORNAT

Allon flu:
Aton
Churton
Coddinton
Farnedon
Clutton

Alderfey
Chowley

Harthill chap:
Buckley
Cholmundelay

Holt cast:
Barton sup mont
Dunkin
Norbury

DENBIGH
PARS.

Dee flu:
Tylston
Shotlych
Worthenbury
Kiddinton
Old cast:
Malpas
Chad chap:

M
Co
Wor

PARS

FLINT PARS.

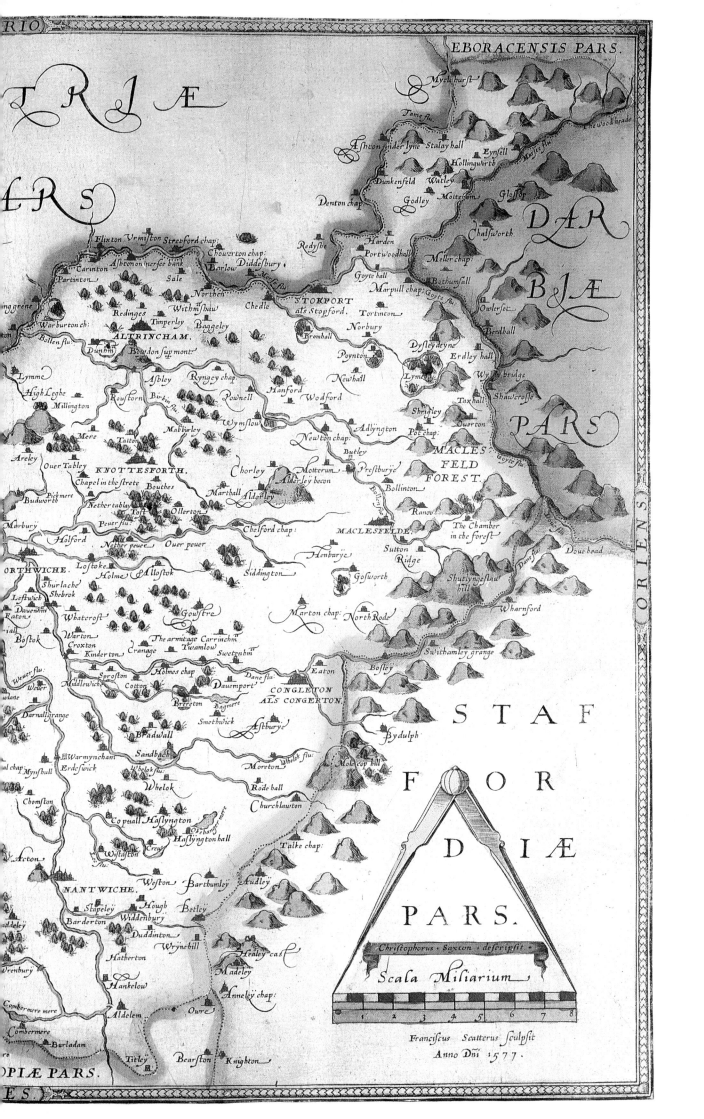

EBORACENSIS PARS.

TRIÆ

PARS

Myelhurst

Tame flu:

Ashton under lyne Stalay hall Eynsell

Hollingworth

Dunkenfeld Watley

Denton chap: Godley Mottersam Gloffor

DAR

BIÆ

Flixton Vrmiston Stretford chap: Redyshe Harden
Carinton Ashton on merfee bank Chowerton chap: Portwoodhall
Partinton Barlow Diddesbury Goyte hall
ng grene Sale Merfe flu: Marpull chap: Goyte flu: Bothumsall
Warburton ch: Northen STOKPORT Tortinton Owlerset
Bollen flu: Redinges Withinshaw Chedle als Stopford Norbury Berdhall
Lymme Timperley Baggeley Bromhall Dysleydeyne
ALTRINCHAM. Poynton Erdley hall
High Leghe Dunham Bowdon fup mont: Newhall Lymerus Wy . . bridge
Millington Ashley Ryngey chap: Hanford Taxhall Shawcrosse
Roustorn Birkin flu: Pownell Wodford Shrigley Ouerton
Mere Mabburley Wymslow Newton chap: Adlyngton Pot chap: MACLES
Areley Tatton Butley FELD
Ouer Tabley KNOTTESFORTH. Chorley Motterum Prestburye FOREST.
Pickmere Chapel in the strete Bouthes Alderley becon Bollinton
Budworth Nether tablay Marthall Alderley Ranoo
Marbury Toft Ollerton Chelford chap: The Chamber
Holford Nether peuer Ouer peuer MACLESFELDE in the foreſt
ORTHWICHE. Loftoke Henburye Sutton Doue head
Shurlache Holme Allostok Siddington Ridge
Leftwich Shebrok Gosworth Shutlyngeslau
Dauenhm Whatcroft Gowstre Marton chap: North Rode hill Wharnford
Eaton Bostok Warton The armitage Carrinchm Swithamley grange
Croxton Cranage Twamlow Swetenhm
Kinderton Holmes chap: Dane flu: Eaton Bosley
Sproston Dauemport CONGLETON
Middlewiche Cotton Bagmere ALS CONGERTON.
Weuer flu: Brereton STAF
Weuer Smethwick Astburye Bydulph
Darnallgrange Bradwall FOR
ElWarmyncham Sandbach DIÆ
Mynshull Erdeswick Whelok flu: Moreton Whelok flu: Molecop hill
Chomston Whelok Rode hall PARS.
Copuall Haslyngton Churchlawton
Wistaston Crew Haslyngton hall Talke chap: Christophorus Saxton defcripfit
Acton flu: Scala Miliarium
NANTWICHE. Weston Barthumley Audley
Stapeley Hough Betley
Barderton Widdenbury
Duddinton Wrynehill
Hatherton Healey caſt
Hankelow Madeley Franciſcus Scatterus fculpfit
Anneley chap: Anno Dñi 1577.

Combermere mere Aldelem Owre
Combermere Burladam Titley Bearston Knighton

OPIÆ PARS.
ES.

ORIENS

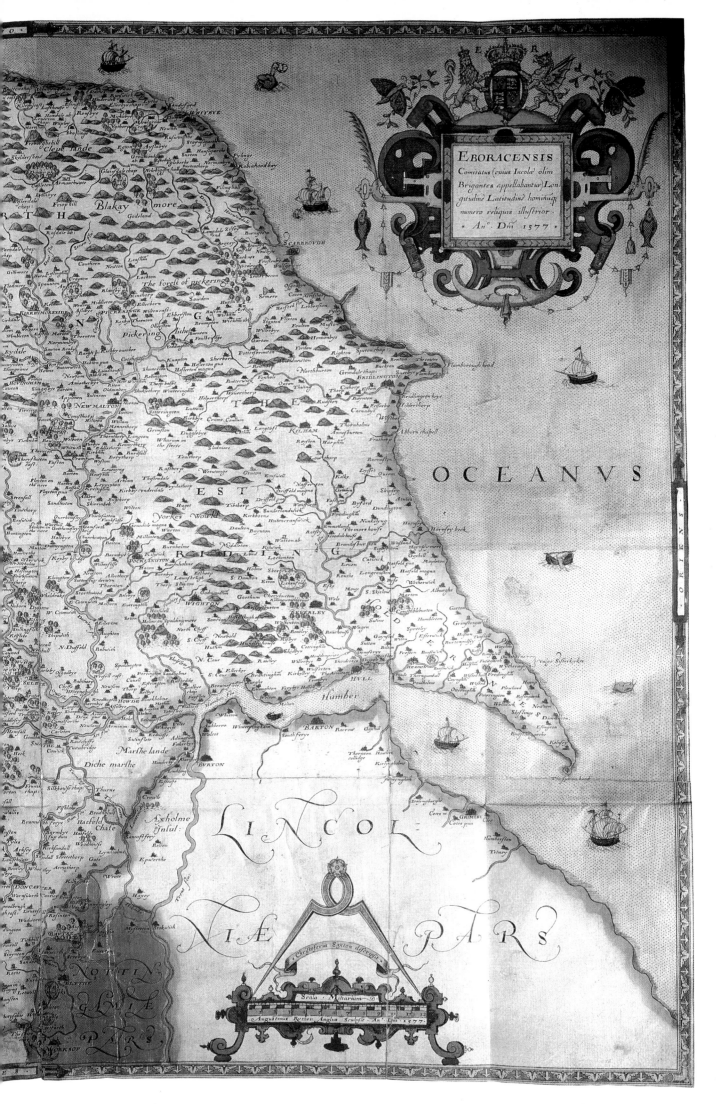

CVMBERLANDIÆ

PARS

WCSTM diæ

KENDALE

Par

OCCIDENS

OCEANVS

INDVSTRIA NATVRAM ORNAT

Scala Miliarium

Christophorus Saxton descripsit

Remigius Hogenbergius sculpsit

CES

MEL

78

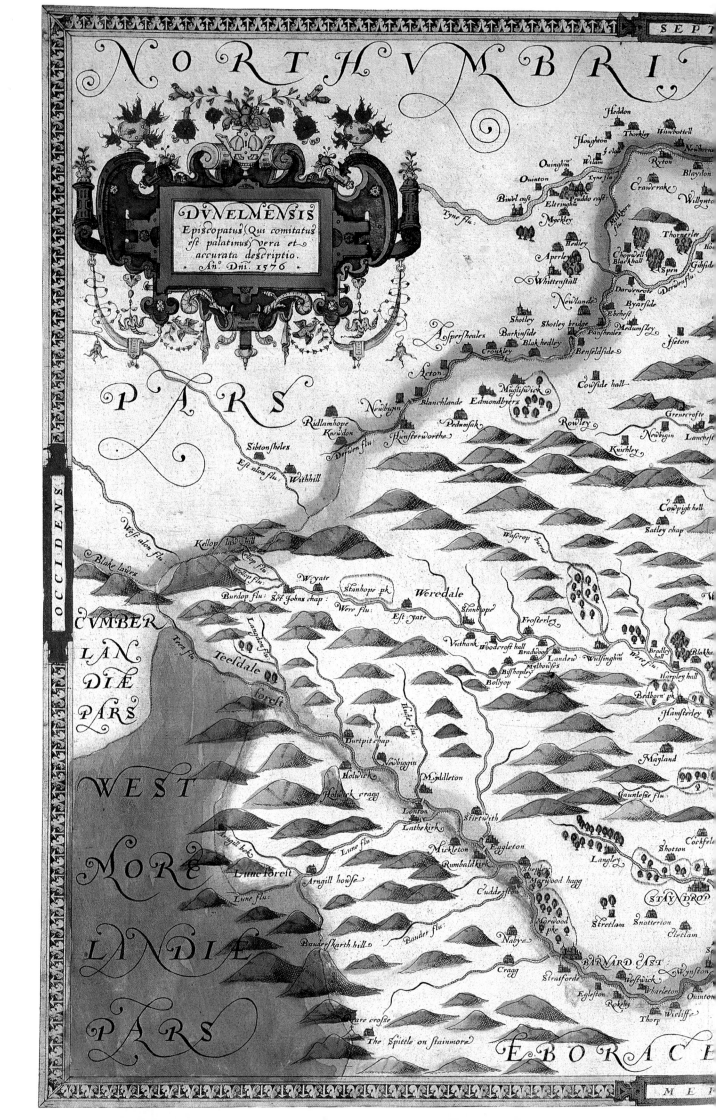

NORTHVMBRI

DVNELMENSIS
Episcopatus (Qui comitatus
est palatinus) vera et
accurata descriptio.
Anº Dni. 1576

PARS

CVMBER
LAN
DIÆ
PARS

WEST

MORE

LANDIÆ

PARS

Heddon
Houghton · Thorkley Winbottell
Ouinghm̄ Wilam · Newhoriea
Ouinton Tyne flu: Ryton
Biwel cast Pruddo cast Blaydon
Eltringhm Crawcrake
Myckley Millgru flu: Willynto
Hedley Thornerley
Aperley Chopwell
Whittenstall Blackhall Aspen Gibside
Darwencote Derwen flu:
Newlande Byarside
Shotley Ebchest
Aspersheales Shotley bridge Medumsley
Barkinside Pansheales Iseton
Cronkley Blak hedley Benfeldside
Acton Cowside hall
Miglifwick
Newbigin Blanchlande Edmondbyers Grencrofte
Ridlamhope Pedemsak Rowley Newbigin Lanchest
Knowdon Hunsterworthe Knitchley
Sibtonsheles Derwen flu:
Eston flu: Withhill Cowpigh hell
Satley chap
West alon flu: Waskrop burial
Kellop law hill
Blake lawes Kellop flu: Wiyate Weredale W
Walſop Flu: Stanhope pk
Burdop flu: Sct Johns chap: Stanhope
Were flu: Est yate Frosterley
Teeſ flu: Langlant flu: Vnthank Woodcroft hall Bradley
Teesdale Bradwod Landew Wulsinghm hall Blakha
Mylhouſes Wore flu:
forest Biſshopley Harpley hall
Hide flu: Bollyop Bedborn pk
Hamsterley
Durtpit chap
Newbiggin Mayland
Holwick
Holwick cragg Myddleton Gaunlesse flu:
Louton Shotton Cockfele
Lathekirk Stirtwith Langley
Arngill beke Lune flu: Mickleton Eggleton
Lune forest Rumbaldkirk Shetley STAYNDROP
Arngill house Cuddestton Marwood hagg
Lune flus: Marwood Stretlam Snotterion
pke Cletlam
Baudeſkarth hill Bauder flu: Nabye BARNARD CAST: S
Cragg Strasforde Westwick Wynston
Eggleston Wharleton
Rokelby Quinton
dure croſſe Thorp Wicliffe
The Spittle on stainmore EBORACE

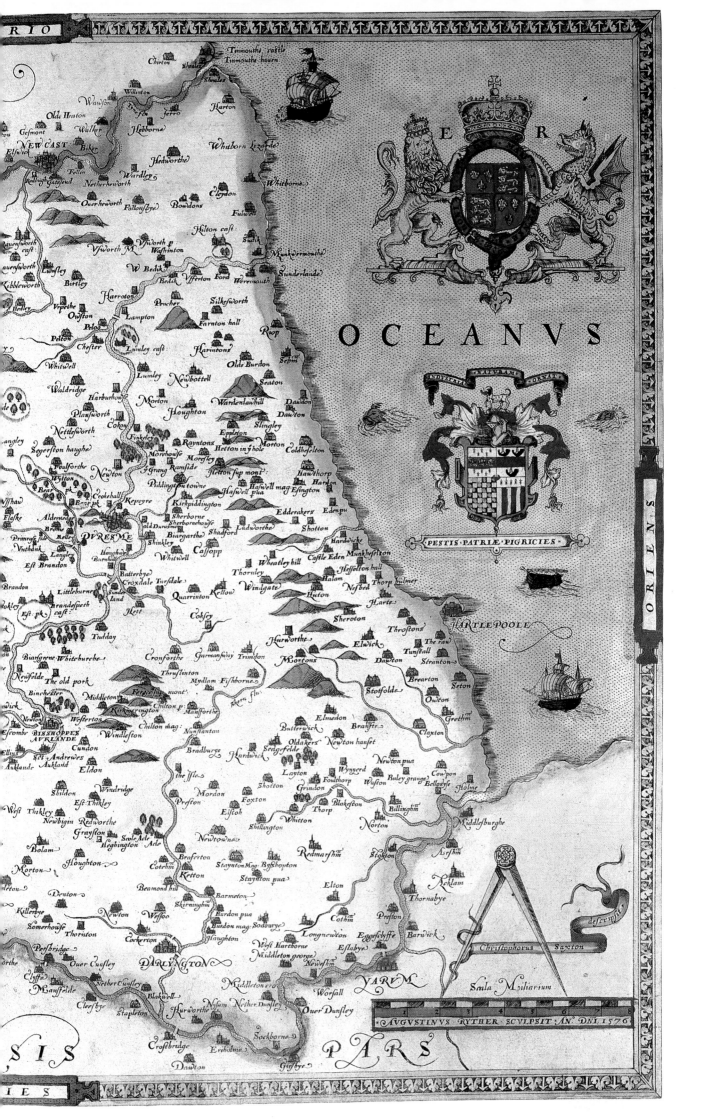

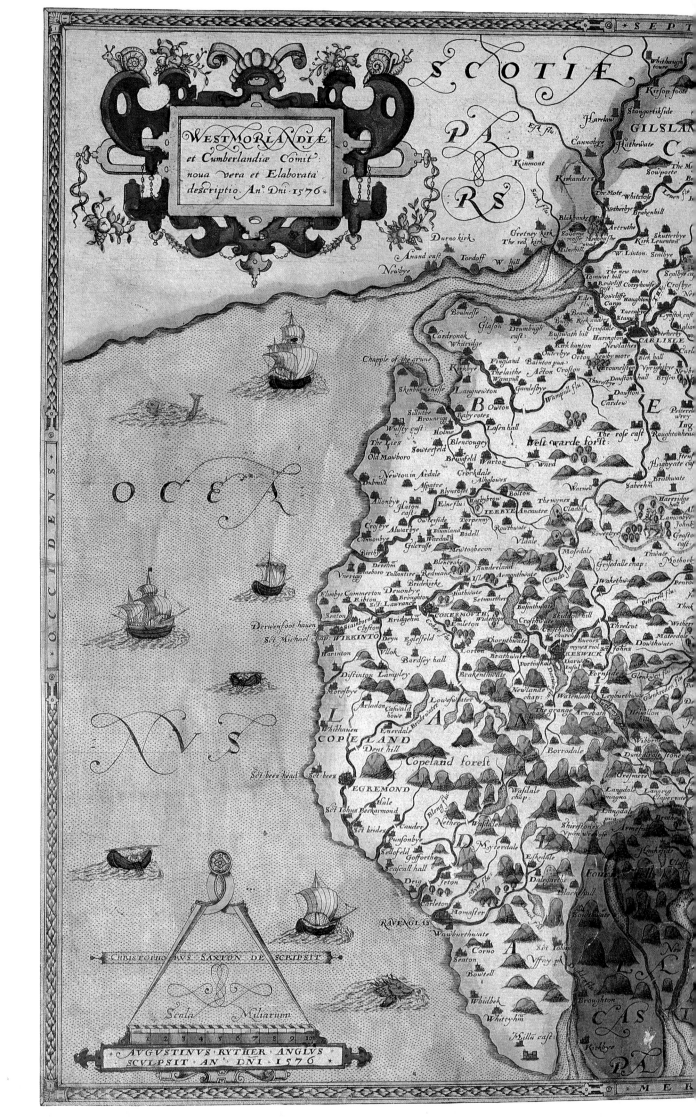

WESTMORLANDIÆ
et Cumberlandiæ Comit:
noua vera et Elaborata
descriptio. An: Dni 1576

SCOTIÆ

PARS

GILSLAN

CARLISLE

OCEANVS

Chapple of the grune

Boulnesse
Glasson
Drumbugh
cast

Cardronok
Whitridge
Kirkbye

Skinburnenesse
Langnewton

Silloth
Brounrig
Raby cotes
BOUTON

Wulsty cast
Holme
Lason hall

The Lies
Sowterfeld
Blencougey
West warde forst:

Old Mawboro
Brumfeld
Warton

Newton in Avdale
Crookdale
Alhalowes

Dubmill
Aspatre
Blynosest
Bolton

Allonbye
Haton
Elne flu
Hathybrow
The mynes

cast

Crofsby
Owteyside
Torpenny
IERBYE
Ancautre
Cladbek

Cannonbye
Alwarbye
Plumland
Bodell
Rowthwate

Birth
Wardall
Vidale

Dereham
Gilcrosse
Mewtooabecon
Mosedale

Vnerigg
Tallantire
Redmane
Sunderland
Armanthwate

Elnaboro
Bridekirke
Cauda flu

Elenbye
Cammerton
Devonbye
Isell
Hathwate

Ribton
Broughton
Setmurther

Seaton
Set: Lawrance
COKERMOUTH
Basinthwate

Stainborne
Bridgehm
Emleton
Widehope
Crosthwate

Set: Michael chap
WIRKINTO
Deyn
Eglesfeld
Thorynthwate

Hartinton
Vilok
Lorton
Brathwate
KESWICK

Distinton
Lampley
Bardsey hall
Brakenthwate

Moresbye
Newlande
chap:
Watenlath

Arladon
Caswald
howe
Lowefwater

Whithauen
Brodewater
The grange
Armebath

Enerdale
COPELAND
Dent hill
Copeland forest
Borrodale

Set: bees head
Set: bees
Wafdale
chap:

EGREMOND
Blenq flu
Langdale
magna

Set: Iohns Beckarmond
Nether Wafdale
Langrig

Hale
Langdale
pua

Cauder
Shirstones
Vpon wodose

Punsonbye
Myterdale

Sellofeld
Gofforthe
Eskedale
Dalegarthe

Seascall hall
Drig
Irton
Black hall

Carleton
Momaster

RAVENGLAS
Wawburthwate

Corno

Seaton
Set: Iohns
Vffay

Bowtell

Whidbek
Whityhm

Halla cast

FURN

CAS

PARS

OCCIDENS

NVS

Scala   Miliarium

CHRISTOPHORVS SAXTON DESCRIPSIT

AVGVSTINVS RYTHER ANGLVS
SCVLPSIT AN: DNI 1576

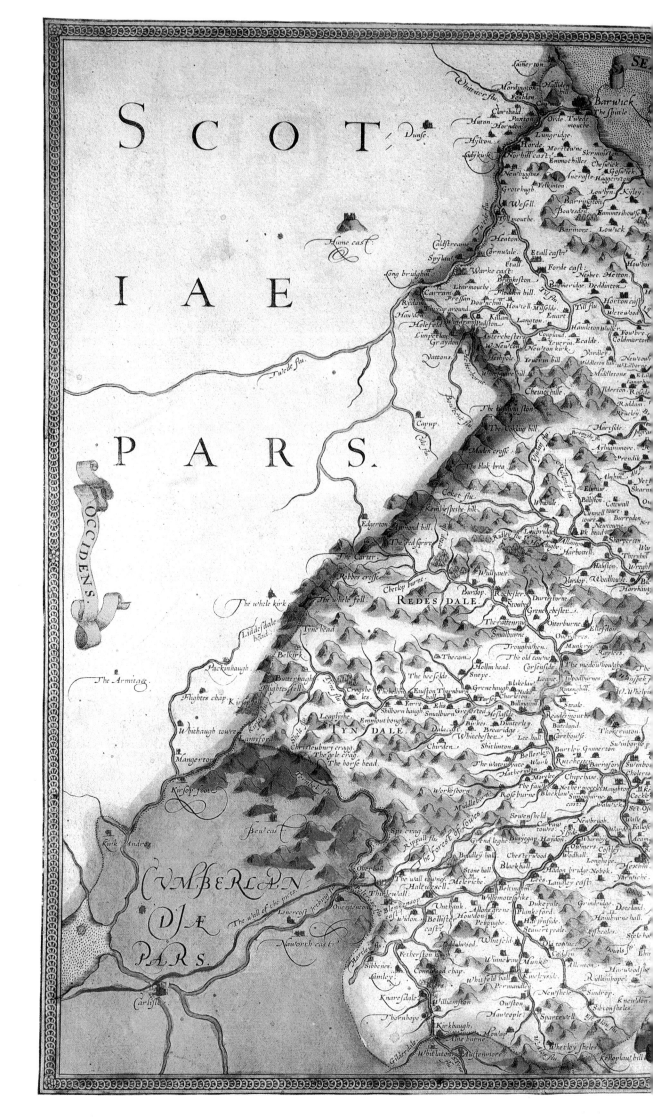

SCOT·
IAE
PARS.

OCCIDENS.

SEA

Barwick
The shiile

Lainerton
Mordington Halliden
Fambton
Whitate flu. Cay Orde Twede mouthe
Claribald Paxton
Huton Horndon Norhill east
Dunse Hylton Horde Moretowne Skrimmon
Gulykirk Emmothilles Cheswick
Newbiggins Increste Haggerston
Grotehugh Felkinton Lowlyn Kyley
Wesell Barrington
Bowesden Sammeshouse
Barmore Lowick

Hume cast. Heaton Howbor
Caldstreame Cornwale. Etall cast.
Spylaw Etall
Long brudghn Warke cast. Forde cast. Nesbet Hetton.
Ltarmouthe. Brankeston Bromeridge. Deddinton.
Carram Pressan Floddon hill Til flu.
Rydam Dounchm. Howtell. Milfelde Till flu Werewood
Haudon thirlp ground. Killam Langton Enart Horton cast
Holefeld Windram Paston. Coupland Hamleton waller Fowber.
Limpeelan Interchester W. Newton Yeuerin. Ecalde Coldmarton
Graydon Hethpole Newton kirk Yardley Newton
Yattons Yeuerin hill Middleton hall W. Lilborne
Whitsquire hill Mikleton Lanneh
Chewiothille Alderton Roser
The touching ston Roddan Reueley
Capup The Coklaw hill Hartside Ingra
Maiden crosse. Aylianmore
The blak brea Prendik
Coket flu. Vnthank Alnham Yett
Kemblespethe. hill Whitsde Elishaw Skarn
Billiston Cotewall
Edgerton Annond hill Clennell toure Burrodon
The red scrue toure Newtowne Ner
The Carter Ridley flu Lynbridge Pk head Sharperton
Robbes crosse Allanton War
Chetlop burne Harbottell Thornhil
The whele kirk REDESDALE Whittaule Halston Wroghl
The whele fell Burdop Rochester Yardop Woodhouse
Liddesdale head. The rattenraw Durtreborne
Tyne head Smalburne Grene cheslev
The Carter Troughthen. The old towne Oiterburne
Packinhaugh Belkirk Thecam Corsenfide Elkston
Butterhaugh The hee felde Hollin head Ouerstres. Munkrys Raylees
Flightes chap Kirson fell Flightes fell Sneype. Leuene Woodburnes The meadowhoughe
The Armitage Ewne flu. The bellys Euston Blakelau Kinsghm. Whelpin
Flightes chap Kirson fell Croyshe les. Thornburn Grenehaugh Nuke Whelpin
Whithaugh toure Yarro Elisa hall Tarset Charleton Billington Streale
Leaplyshe Shilborn haugh Smalburn Grested Heslafide Reaßenmouth
Lanyford Emmout bough Birkes Dunterley Burdland
Christeubnry cragg Dalacost Breavidge Lee hall Grehowse Thokervinton
Mangerton The gele crag Whitchester Churden Shitlinton Bellerby Burtley Gromerton Swinborhe
The horse head The water-yate Wauk Rutchester Barrsford Swinbo
Hathern-ton Morelee Chipchase Cholerto
Kirsoy foot Workesborn the faule Nether moreles Haughton M.K
Roseburne Blacklaw Simonburne Cockle
Beucast Middleborn Walwick Set Os.
Sewenfield Walle Fallow
Spu eirg Carraw Newbrugh Sde
Rippall flu toure
The Forest of Lowes Grend leghe Busyegay Haydon Wharnby Costley
Kirk Andres Bradley hall Chesterwood Wodhall Longhope
Ouerhall Stone hall Blackhall Hadon bridge Nobok Hexbm Yarnbche
The wall towne. Lees Lanley east
CVMBERLAN Halturfell Melriche Beltingbm
DIAE The wall of the picts Thurlewall Willymoteswike Dukepule Grindridge Dotelan.
Lenercoft Ieching in Ouerstenton Blenkensop Unthank Allanegrene Plankeford Heinburne hall
PARS. widon Bellyster Houdon Penpughe Haltenside Stauert peale offheales. Steele ba
cast. Toddalwood. Whitfeld Caldon Douols Elin.
Naworth cast. Harties flu. Fetherston haugh Winnewra Munk Alenton Hawwodsd.
Sibbenes. Comewood chap. Whitfeld hall Kneleyside Newshele Riddlanhope
Imley Permandley Sindrop Knowton
Carlisle Knaresdale Williamston Ouston Spartewell Sibtonsheles.
Thornhope Hawteople
Kirkhaugh Moular
Gilderdale Whitlaton Austenmore. Whetley sheles Kesloplauf hill

84

NORTHVMBRIÆ
COMITATVS
(Scotiæ contiguæ)
Noua Veraq̄
deſcriptio.

ORIENS.

OCEANVS.

DVNEL
MENSIS
EPISCOPA
TVS PARS.

CHRISTOPHORVS SAXTON DESCRIPSIT

Scala Miliarium.

MONVMETHENSIS
Comitatus Regis
Henrici quinti
natalitijs celeberrimus

An° Dm̄ 1577

BRECKNOK
PARS

INDVSTRIA·NATVRAM·ORNAT

GLAMORGAN

PARS

· Christophorus · Saxton · descripsit ·

Scala Miliarium

1 2 3 4 5 6 7 8

OCCIDENS

Llantony
Honddu flu
Walterston
The Fathe
HERFORDIE
PARS
Theolde
Cwmore
Tre
Pentriffo
Stanton
chap
Michaelchurch
Cruçorne
Skiriduauwe hill
Groynor flu
Bettus chap
Peterchurch
Pentrieielhill
Irrlay
Landihgarthen
Hentheye
Llangeny
Warnde
The chapel
Llangryney
ABERGEVENNY
Warnigoch chap
Llaneler
Llanwenarth
Llanforst
Colbro
Llanth
Blorench
hill
Hardwick chap
Llanfihangel iuxta
Llanhileth
Llanguerif
Awon flu
Blanagwent
Capel newith
Gortre
Elweth uachfu
Llanyhangel
pontemoiel
Teethwen
Llankyrthe
Panteage
Rompny flu
Srowanflu
Capelbrathiere
Bydwelthy
Terestent
Llantheruell
chap
Llamureg
Llayrhange
tonnoyn
Tumberlow hill
Monethuslayne
Heullis
Malpas
Ryfeler
Bettus chap
Kethdygaier
Kirten becon
Crindye
Rocheston cast
Bedwel
NEWPORT
Maughen
Grenefeld cast
Caierfilee cast
Ridway
Bassaleg
Gwarnea lena
Elwith flu
Tredeager
Llannyhangel
Veddo
Sct Bryde
Keuenmahler
Coydkirne
Marshfeld
Llanedevn
Sct Melens
Peterston
Lloynigrant
Rompney
Rothe
Thefplatt
CARDYF

HEREFORDIÆ PARS

Kentchurche
Treueraren
Garwey

GLOCESTRIÆ

ILannoyth
Skenfrith caft
Norton
Whitchurch
Welfhetow
Gennareou
Monmentensis Pars
Wolfhe hecknor

smond caft
annaier
caft

ILanruthall
Englyfhe hicknor
Welfhetow

Sct. Moughan
Perthire
ILamuaner chap:
ILangattock
Vibouauel
Rockefelde
Sct. Michaell
Dixton

Scythery ffenny
ILanyfhangel
MONMOUTH
Tra:
Stanton

Gracedue
Wonaftow
Newland
Peurefe
Graowey
Pennalth
arthe
Treargare
Dingeftow
Micheltroy

Breynywin
Penclawithe
Helenwoodehall
ethey
Raglan caft:
Penclafe
Trylegh
Sct. Breuels

nevath
Olu ze flu:
Langouen
Landogo
Brockwere
oftrey
Llandenye
Lanyffen
Trileghgrange

Gwerneffeyre
Llanfoye
ILanmihangel
torremouyth
Ynterie pux:
Tintern ab.

Llangeyre
Chapelhill
Sabrina flu:
VSKE
Llangomes
Wolfenewteu
Parcaftl:

adoel
Kilgeruck
Pentorye chap:
ILanllowell
Newchurche
Scr. Arian
Cincante chap:

egeby
The anane
eth
Llantriffen
Itton
Houock
Irdnam

Strogle caft:
CHEPSTOV
redonock
Enterwoode
Threggy flu:
Shirenewton
Mounton
chap:

Kemis
Marthellre chap
Dynham
Ruyfton
Mathern

eueffeuord
Creke
Scr. Pere
Beitefley

neuock
Llandeuaigo
 Lanuaier
anughes
Caierwent
Porefkewet
Scr. Trenele chap
Auft

The flu
Llanbed
Scr. Bryde
Rogeat
Caldycott
Sudbrok

riftchurche
Penhow
Vton
Treinetie chap:
Pennegoyd
Llannehangell
Langfton
Llanmerten
Woudre
Chaifton rock

Llanwarn
Briffheton caft:
Magor
Chefel pill

e MORE
Redwicke
Sabrinia flu:

Wrtfton

Goldecliffe
PARS

Goldeclyffe poynt
The ilennye Iland
Kinges rode
Porfhutpoynt

Auon flu

BRISTOL

SOMERSETENSIS
PARS

ORIENS

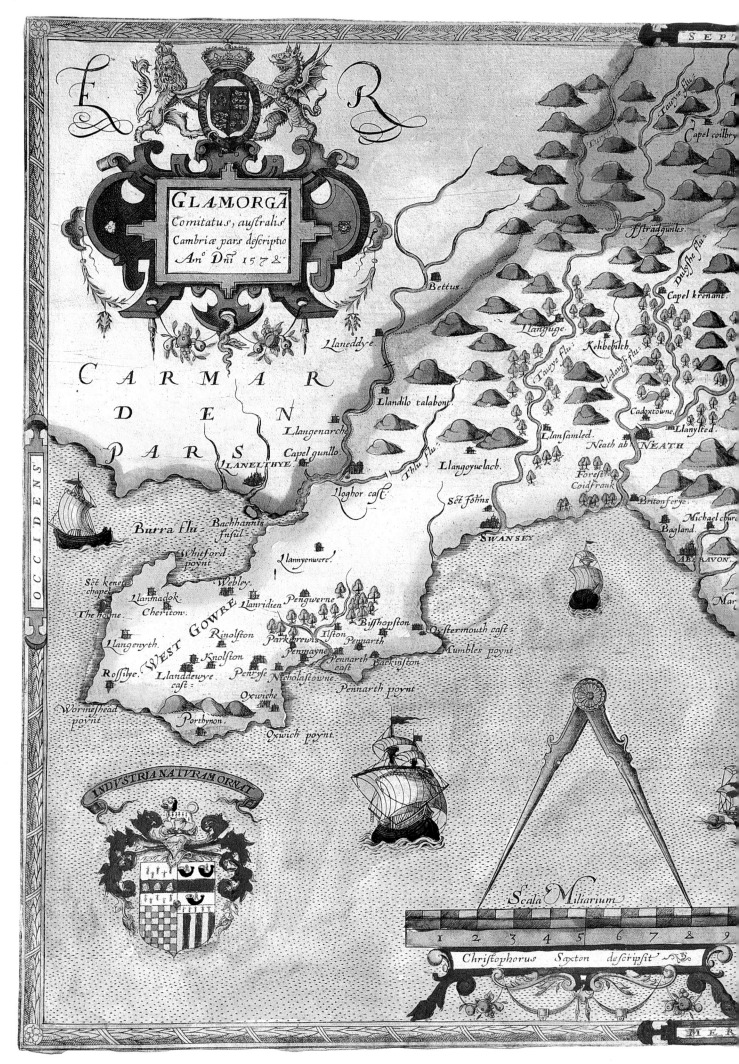

SABRINA flu:

Whire hall
Teme flu: Beitus

Simon hill
Sabrina flu: Llanbadernueny bha Begildye Llanuarewaterden Weston
Tye flu: Skibborray. Stou
Llanydlos Caft Dynbod The forest of Knukles. Knukles Teme flu:
Llangeryk Llannano & Hyop Tandij Branton
Tye flu: Chanpyter Llangunllo Monachtree Knighton bryan
Darnolflu: Weftonhall Pillethe Lug flu:
Combeyfroythe Whitton Norton Llyngan
Sct: Harmon Combehyre Bleidyogh stepleton
Eland flu: Dulas flu: Llandiwye The forest of Hereford Fulalbro
Bletuugh Lliton Cafcop pars. Dicoyde Prestayne
Llanfanfraid Clawedy flu: Llanyhangle redithan Ednall New caft
Rayader gowy Nantmel Comaron flu: Radnor HERE
Garthuagh Llanbadern forelt Kinerton Angoy
Llyngwyn Llanyhangle rehgan Llandeuby Cueuilles Somergil flu: Knill FOR
Clarwen flu: Llanner wye flu: Caft Cueuilles New Radnor Kyneton DIÆ
Llanutbel Llanyhangle Harton Hargaft
Llanadrindod nantmelan Old Radnor Arro flu:
Lleftinan Hawye flu: Beitus chap. Llanhangle arro Huntingto
Ketbualgarth Dyffart Llanfanfraid Edwar flu: Gunchel flu: Glostre Colwey chap caft
Capel Llanhangle Colwin caft Tregrend Glafcomb
Delas flu: prinpabon Glascomb Newchurch
Llanyhangle abergwessen Weuery flu: Llanauon uawre Llanolway Rulen Brynguyn Llanhangle arro Bryllay.
Llantheuye abergwessen Llanauon Llanuareth Llanbadern Machway flu: Whitney.
uawre Bealt Abcredway Caft Beitus
Llanganten Maylmyns Llynbogblen Payne
Llanquan Llanynys Llanthewye comb Llanbedder Llandewye Clyfford caft. Wye flu:
vachan Portheroyes Altemawre Llanuilo chap: Clero
Llanllowenuel Capel Cunok Hay Kewfop PARS.
Yruon flu: Gwendor Llowas Duleflu:
Capel pylyn Llangamarche Capel defrune Crecaderne Llanftephan Glaiebury Llnygon Capelbrengoran Crafwel.
Iftrodefyne honthye Boughrud wye flu:
BREK N O Llangoyd Aberlleuenye
Capel newith Honthy flu: Llandeuathley Llyfwen Pipton
Llanbrayne Merterkynok Llanyhangle uachan Porthamble Capel a fyne
Llanydiloruayne Ifker flu: Broynllys caft Talgarth Hothny flu:
Llywel Trecaft K E Talachye Llanuyllo Caft Dynas. Llantony
Iftradwalter Gwetherik flu: Garthbrenaye Llanduialog Llangouilog The fothok Comyoye
Llanuaterarbryn Heuud pk Traftong Battel chap Llanthew caft. Llanyhagle Llangors Herefordie pars
Llanymthefry Capel ridbrue Aberyftory Breknoke tallyn Llyn Hothny flu:
Aberbrayne Llanspythed Newton Quachant Ketheden Pertriffo
Ufke flu: Capel beitus Aberkinurik Sketliog Llanwarn Llangiftye. Llanyhangle
Muhuy flu: Deuinok Comlas flu: Cantreffe Blaynllynuye carndye,
Llanyhangle Cray flu: Touremaltwalbury Penuch caft Llanfanfrayd
muthuey Capel yllfyr Llanuranach Tretowre Peterchurch
Cledagh flu: Llanuygan Gilfton Llangenye Crecowell
Sourhey flu: Capel fenny Monuchdennye Llangonder Ufk flu: Llanwenarth.
Llanthewyefant hil Llandettye Llangroyney Llangattok
The black Neath flu: Llanelye Abergeuenny
mountayne Meley flu: Llangattok
Capel Trawgarth flu: Taue uachan flu:
Llanthewyefant Callweit Hepfey flu: Capel taueechan
R Tau flu: Iftradwelthye Taue uawre flu: Capel nantye
Capel Gaynor Blanagwent
Coyelbryn Neath vachan flu:
Iftradgunlas Penederyn Morlafhe caft Rumpey flu: MONVME
THENSIS
Aberpirgum Abardaier PARS.
GLA MORGAN
Neath flu:

Chriftoferus defcripfit
Saxton
Scala Miliarium
1 2 3 4 5 6 7 8 9 10

PEN-BROK
comitat qui inter meridionales
cambriæ ptes hodie ceñsetur olim
demetia L Dyfet B hoc est
occidentalis walia descriptio
Anⁿ Dñi 1572

OCCIDENS

MARE

HIBERNICVM

Langlas head
Strumble heade
Thecowã Calfe rock
Llanunda
Manernowen
Fisc
Set Nicholas
Llannast
Set Katerens
Treuenyth
Merther
Llanyran
Vttorstowne
Llanhowel
Llanrithon
Llanedryn
Set Lawr
Set Dauids head
Goridchapel
Llandeloy
Whitland bay
St Dauides
Haies east
Whitchurch
Riefton
Brodye
Set Stenans
Set Ayluew
Ira
The byffhop and his folaixes
Set Nones
Doluuth hauen
Plumfton rock
Porclais
Shinlaterdek
The byffhop
Ramsey Insul
The hoyfe
Roohescaft
Canr
Rihyerock
Knowleton
Lamfton
The lytel hauer
Harreston
Goltop rode
West Walton
Set
Set Brides Insul
Talbenyre
Walwin caft
E
Hafcard
Robertfton
Ypri
INDVSTRIA NATVRAM ORRNAT
Sandshaue Neffton
Gellyfick
Yardlanftone
Marlas
Harbayfton
Hubrst
The Scaline Insul
Set Annels
Sandy hauen
Grefholme Insul
Midlan Insul
Gatehoine Insul
Dale
Dale rock
Ytack rock
Mowftone
Theblockhowse
Set Anchap
Ladye chap
Stockeholm Insul
Nangle
The blockhouse
K
Milfordhauen Radnor
Shepe Freslie
Freshwat
Gu

CARDIGAN
PARS

Penkemas poynt

Tuy flu. Cardigan
Llangordmore

Sct. Dogmels
Oscoidmortuner

Reuen *Capel Laughrid*
brynn

Egloswither

Treuethel *Kilgarrencast.*

Llantwood *Manerdue* *Kenarth*

Brydeithe *Capel Kiluaure*

Bayuil *Llanyhangel penbedo*

Neuerne *Capel colman*

Eglofserrow *Cap: euan*

Maenay *Cap: Nantgwin* *Penrith*

Newport *Neuern flu.* *Kilryden*

Kilgwin *Whitchurch Cap: kestellan* *Cledye*

Wrenny uaur hil

Llanneallordag *Cap: bettus*

Gwyne flu. *Clethy flu.* *Llanuurnach*

Pontuaine

Moruil MAR *Percelly hil*

Monaghlogdee

Caftel male *Caftel henyye* *Llandilo* *Llangludwen*

Caftlebighe THEN *Llangolman*

newcaft: Mancloghay *Kilymanelleyd*

gwels *The mote*

Reuelfton Ambloston *Lifurayne* *Llannkeuen*

Walton *Bletherfton* PARS

Spittle *Clarbafton* *Tansilio*

Egremond *Caft: deram*

edbaxton *Llandewre*

Wifton *Llanhaddon* *Llanbeder uolfray*

Sct. Leonarder chap: *Robefton Creno*

Preadedgeft *Slebaele* *clethy flu. Cannafton* *Narbaith eaft:*

Pickfton *Sifterhowfes* *Kilfane* *Narbarthforeft*

Ofmafton *eaft:* *Redcaft*

Harreld *Mynware Newton* *Munkton Caft: moherry*

fton *chap:* *Templeton* *Crunmier*

Boulfton *Martheltwy* *Llewchurch* *Marroe*

Frefthorp *Drinnafton* *Amriath* *Earware*

Rifecaft *Renalfton Coidrath foreft*

poynt *Coidkenles* *Loueflon* *Begely*

Hangum *Lawrenny Crefwel Jeffrayfton* *Sct. Iffels*

Benton eaft:

Burten *Villiamftonpke* *Williamfton chap:*

Roofmarket *Redbart*

Kingwode *Munkefton rock*

Lanftadwel

Upton Carew eaft: Tenbye
Collenfton

Nafhe *Gunfrefton*

Paterchurch Penbroke *Sct: Katerens*
Eft Penner *Sct: Florence* *Giftor poynt*

Burlaxton *Lanftay court* *Trelloyne Pennoly*

Thefte Munkton Lanfaye *Jamefton Northard*

Capel danyels Trewent *Marberbury Ludfop*

Orielton *Hoggefton* *Caldey Infül*

Kerikmahren *Freshwater* *Ludfop poynt*

Sct: Pattriarke

Waren Sct Twinels

Merian

Leuafton Stackpole *Stackpoole orde*
South Carew Broffhefton *Brode hauen*

Kerikmalo

Newton *Sct Gouens poynt*
Sct Gouen

Scala Miliarium

| 1 | 2 | 3 | 4 | 5 | 6 | 7 | 8 | 9 | 10 |

Chriftoferus Saxton deferipfit

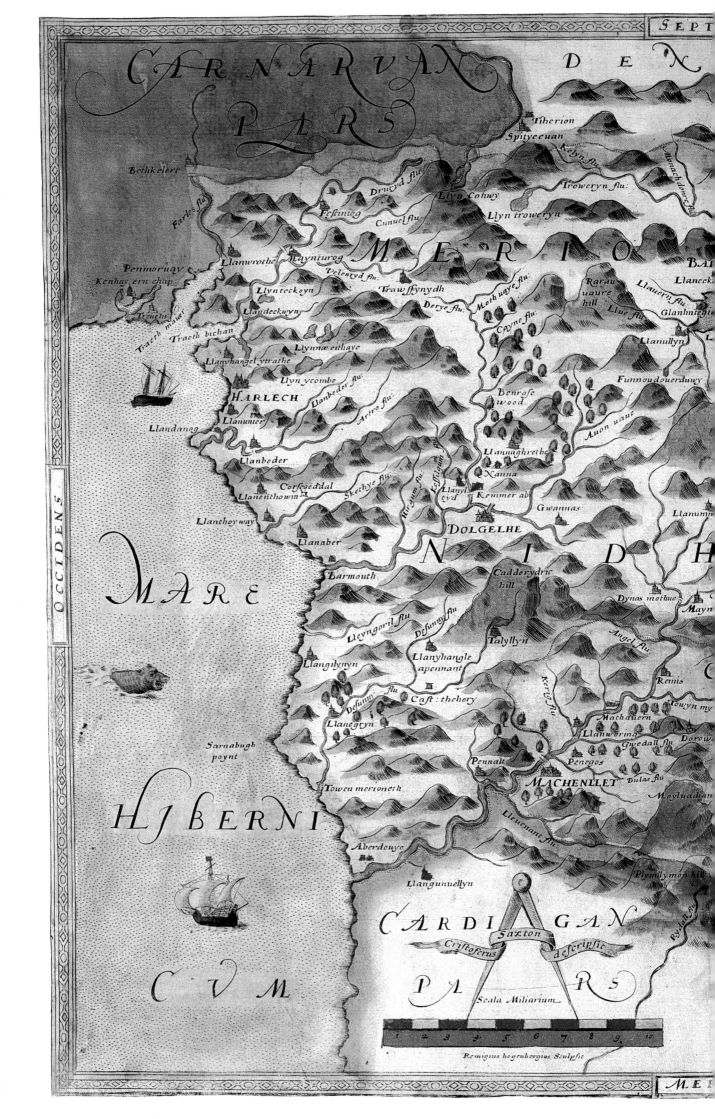

SEPT

CARNARVAN PARS

DEN

OCCIDENS

Bethkelert

Penmoruay
Kenhav ern chap

Traelis flu

Farles flu

Treaeth mawr

Traeth bichan

Llandaneg

Llanwrothe Maynturog

Llynteckeyn

Llandeckuyn

Llynnæ eithave

Llanyhangel ytrathe

Ulyn ycombe

HARLECH

Llanunier

Llanbeder

Corsgaeddal

Llanenthowin

Llanthoyway

Llanaber

Barmouth

Lleyngoril flu

Llangilynyn

Desunny flu

Llanegryn

Sarnabugh
poynt

Toweu merioneth

Aberdouye

Llangunuellyn

Festiniog

Druryd flu

Cunuel flu

Llanbeder flu

Artro flu

Skethye flu

Deshnny flu

Caft: thehery

Llanyhangle
apennant

Velenryd flu:

Trawffynydh

Derye flu:

Llyn Cohwy

Tiberion

Spityeeuan

Kelyn flu

Troweryn flu:

Llyn troweryn

MERIO

Mothuaye flu

Cayne flu

Benrose
wood

Llannaghrethe

Xanna

Llanyl
tyd

Kemmer ab

Gwannas

DOLGELHE

Caddorydric
hill

Talylly n

Pennalt

Lleuenant flu

Plymllymon hill

Rarau
uaure
hill

Llue flu

Funnoudouerduwy

Auon uaue

NID

Dynas mothue

Angel flu

Kerty flu

Mochauern

Llanwrina

Penegos

MACHENLLET

Dulas flu

Gwedall flu

Moyluadian

BA

Llauern flu

Glanhnegie

Llanullyn

Mayn

Remis

Towyn my

Dorow

H

Llanumm

MARE

HIBERNI

CVMM

CARDIAGAN

PARS

Saxton
Criftoferus   defcripfit

Scala Miliarium

1 2 3 4 5 6 7 8 9 10

Remigius hogenbergius Sculpfit

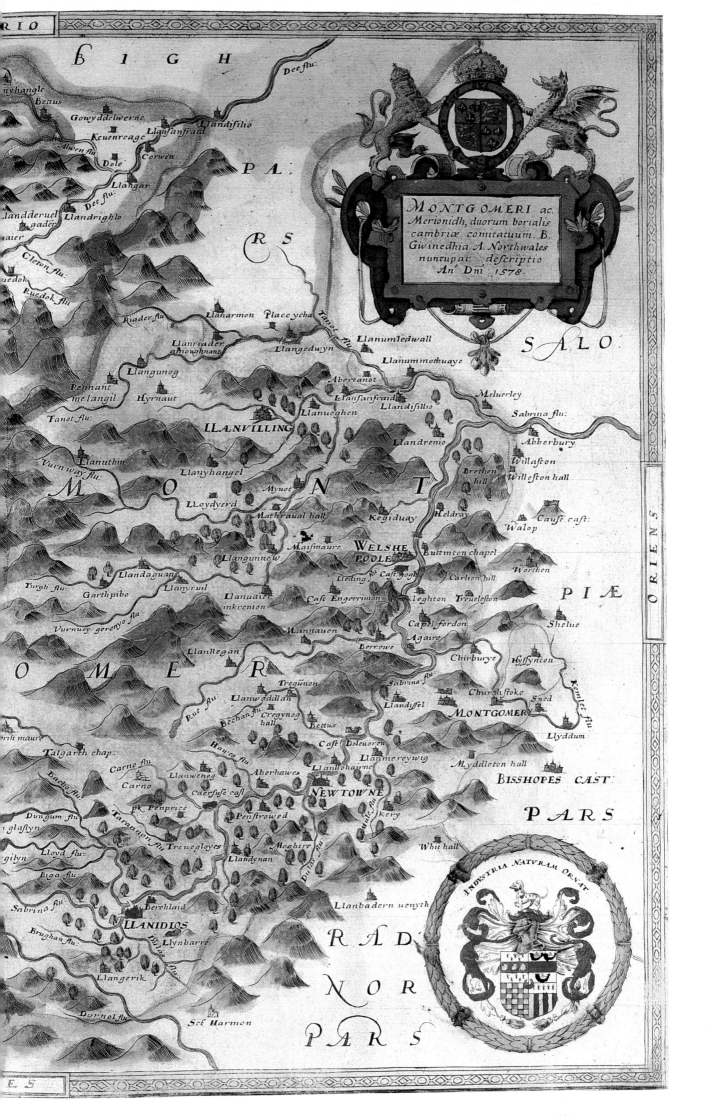

BIGH

Dee flu:

nyhangle
Bettus
Gowyddelwerne
Keuenreage
Llanfanfraid
Llandifilio
Alwen flu:
Dole
Corwen
Llangar
Dee flu:
landderuel
gader
Llandrighlo
maier
Cleron flu:
uedok
Ruedok flu:

PA:

CRS

MONTGOMERI ac.
Merionidh, duorum borialis
cambriæ comitatuum. B.
Gwinedhia A. Northwales
nuncupat descriptio
An° Dni 1578.

SALO

Riader flu:
Llanarmon
Place ycha
Tanot flu:
Llanumledwall
Llanriader
amoughnant
Llangedwyn
Llanummothuaye
Meluerley
Llangunog
Abertanot
Llanfanfradk
Llandifillio
Sabrina flu:
Pennant
melangil
Hyrnaut
Llanueghen
Abberbury
Vurnway flu:
LLANVILLING
Llandrenio
Willaston
Llanuthin
Willeston hall
Llauyhangel
Brethen
hill

MONTI

Myuot
Cause caft:
LLoydyerd
Walop
Mathraual hall
Kegiduay
Heldray
Maifmaure
WELSHE
POOLE
Buttinton chapel
Worthen
Llangunnew
Lleding flu Caft gogh
Carlton hill
Llandaguan
Turgh flu:
Garthpibo
Llanyrul
Llanuater
inkvenion
Caft Engerrimon
leghton
Treueleston

PIÆ

Vurnuey gorenyo flu:
Mannauen
Capel fordon
Shelue
Agaire

OMER

Llanllegan
Berrowe
Chirburye
Hyfsyncon
Tregunon
Sabrina flu
Rue flu:
Llamweddlan
Church ftoke
Sned
Bechan flu:
Cregynog
hall
Llandyfel
MONTGOMERY
rin maure
Bettus
Kemlet flu
Talgarth chap:
Caft Doleuoren
Llyddum
Carno flu:
Llanllohaurne
Llanmereywig
Myddleton hall
Bacho flu:
Carno
Llanweneg
Aberhawes
BISSHOPES CAST:
Dungum flu:
Caerfufe caft
NEWTOWNE
pk Penprice
glaflyn
Taranhon flu Treuegloyes
Penftrowed
Kery
PARS
gilyn
Lloyd flu:
Moghtre
Llandynan
Biga flu:
Whit hall
Sabrina flu:
Berehlaid
Llanbadern uenyth
Brughan flu:
LLANIDIOS

INDUSTRIA NATURAM ORNAT

Dulas flu
Llynbarre
Llangerik

RAD

Dornol flu:
Sci Harmon

NOR

PARS

VIRGIVIVM SIVE HIBERNICVM

MARE

Talackrey
Grouant
Preftatin
Meliden
Gulgraue
Mosfrow
Gwenuskor
Llanafaphe
Treer eaft
Dyfart
Relufnoyde
Whitford
Potruthan
Rudland caft.
Seuton flu.
Combe

Ormeshead poini
Llandrighlo
Llandidno
Penrin
Eglos roße
Llaugwennyn
Llanddlas
Hendray
Abergele

Dafart
Llanfanfraid
Llanelian
Llanshanshore
ABERCONWY

SCT ASAPH
DIFFREN CLWYD
Demyrchion
Maysnynan
Caierwis
SK

CAR

Beitas
Cap. Funhown uaier
Llewenye
Maghegreg
Poguary

Caierheaue
Elwy flu.
Henllan
Snedck
Llandurnog

Meleirgycha flu.
Eglos uagh
Elwy flu.
Foxholes
Whitchurch
Chwyd flu.

Llaubeder Pennyn
Llanufydd
DENBIGH
Llanumnis

NAR
Mananah
Llanuaier
 Mea flu.
Ystrad flu.
Llanrayder

Llynydulyn
Truerite
Llangirnew
Llanfannan
Bachmbid
Plaswar

Llyncoulwid
Llachiged
NANCLIN
Llanuo

Gwider
Llanruff
Barrog
Clawedok flu.
Capel kyffyllyock

Llanrughwzit
Gwetheryn
Vene
Clocanok

Llynyenoneth
Manyan
Chwyd flu.

VAN
Llynaled
Derwen

Lleggue flu.
Capel beitus
Yrton shion
Capel garmon
G

Neag flu.
Llankerigedredion
Thuen flu.
Llanyhangle

Llanpenmaghno
Capel pentreuidog
Holyn
Gyrou flu.
Bettus

Tiherion
Yspity euan
Conwy flu.

PARS
Llangum

Llyn Conwy

Llandderuel gader
Llan

BALA
Dee flu.

XERIONETH

Llyneugnd

Scala Miliarium

1 2 3 4 5 6 7 8 9 10

Chrijtoferus Saxton deferipfit.

PARS

Remigius hegenbergius fculpfit.

MARE
HIBERNICVM
SIVE
VERGIVIVM

Mare Marlroniad

OCCIDENS

ILanbadern
ILanvoderis
ILanvehell
ILannier
ILanflewen
ILanrithlad
ILanathlr
Gronant
ILanuoyr
ILanuigel
ILanallibio
ILanenghenel
Pontrutpont
ILanuaierinarbu
ILawrhangel
ILanualog
Ynys wealt
Cap Talallyn
Aberfraw
ILaighwfen
ILanuerian
Beddon
Malhtrathe
ILanthouwrnw
Amlwich
Gregorfia
ILinoryflu
Afow flui
ILanbal
ILanydiant
Bodedern
Kirghbig
ILaghhull
Treualghmay
ILanbulan
Cap ILeaghgwn
Precoaddluet
ILandrogawn
Haneglos
Cap Tnaier
Beddoveryffa
ILangaffo
Borgwercha
ILanuaieroombe
Treuarthin
ILangynwen
NEWBVRGH
ILanclian
Ynis Ligod
ILinvenllerveg
Capel thon
Dula flui
ILamhangel penrose
ILandorure
ILanalthgo
ILanarghwrn
ILangwythlog
ILanthufnam
ILanbedern
Treggvon
Cyrdana
Hardrauay
Redgint
Hendregadog
ILangrysdolis
ILangrifwen
ILanuaier inhane emye thoy
Pentir eagh
Teumonon
Pooe nem
ILanuaier am pul
girigil
ILanedwen Mincht
ILamdan
Nante

Dinas dynlle
Caer Iere
eurode
Henem flu
Clenogunure

Abermenai
ferre

Meniai fiu

CVIERNARVAN
ILanboblig
Sant flu
ILanuaglan
Sct Elyn
ILanunda
Carrog flui
Gwelye flui
ILandireg
Glyn ILryuan
ILryuan flui
ILanlleueny
Gynhaladoyfen
ILanurlhayrne
ILangaby
Carngwgh chape
Wgynderis
Earoh
ILanstindwye
Piftill
ILanarmon
NEWIN
Edarne
ILanmor
Boduran
Gache flui
ILandidwen
ILanbedr
ILandugwr
ILankean
ILanfangunuehill
ILanyngan
Nangunadle
Botunnog
Saughe flu
Bryncroffe
Boduern
chap
ILanuaielryfe
Darey flui
Rue
Gwelyn Infula
Bardefey Infula

Christophorus Saxton deferipfit

Scala Miliarium
1 2 3 4 5 6 7 8 9 10

Penwenkele
pornt
Brayehrpult
poynt

Mervrofs Infula
Stidwall Infula

IVLLHELY
Carodimil rock

INDESTRIAN ATER